O9-BHK-904

# GUIDE TO THE NMR EMPIRICAL METHOD

## A Workbook

# GUIDE TO THE NMR EMPIRICAL METHOD

## A Workbook

Roy H. Bible, Jr.
G. D. Searle & Company
Chicago, Illinois

Roy H. Bible Jr

PLENUM PRESS · NEW YORK · 1967

Library of Congress Catalog Card No. 66-11695


*A Division of Plenum Publishing Corporation*
*227 West 17 Street, New York, N. Y. 10011*


*Printed in the United States of America*

# PREFACE

The organic chemist who wishes to learn how to use NMR spectra effectively must first learn the essential facts and then must gain both ability and confidence through the solution of a wide range of specific problems. My previous volume, Interpretation of NMR Spectra: An Empirical Approach, was written specifically to present and explain the necessary background material. The present volume is designed to provide the reader with a full range of experience in the interpretation of NMR spectra. The exercises are arranged in a sequence designed for rapid assimilation of not only the basic concepts, but also increasingly more complex details. Emphasis is placed on the difficulties normally encountered in the use of spectra and also on the many practical aids which are helpful in overcoming these difficulties. For most of the problems, at least one reasoning process is outlined by which the questions can be answered.

This text is, in part, an outgrowth of my participation in workshops which were held at Canisius College under the direction of Dr. Herman Szymanski and at the College of Pharmacy of the University of Illinois under the direction of Dr. Charles L. Bell and Dr. Ludwig Bauer. This experience has been of considerable aid in the formulation of this workbook.

Most of the spectra used in this book were obtained by Searle staff members in the course of their own research. I am particularly indebted to the following workers for permission to use their spectra in the preparation of this workbook: Dr. Clarence G. Bergstrom, Edward A. Brown, Dr. Robert R. Burtner, Dr. William E. Coyne, Dr. James R. Deason, Robert W. Hamilton, Dr. Willard M. Hoehn, Dr. William F. Johns, Dr. Max J. Kalm, Dr. Stephen Kraychy, Ivar Laos, Dr. Ernest F. LeVon, Dr. Calvin H. Lovell, Dr. Joseph S. Mihina, Robert T. Nicholson, Dr. Viktor Papesch, Dr. Richard A. Robinson, James M. Schlatter, William M. Selby, Dr. Paul B. Sollman, Bruce G. Smith, Dr. Robert C. Tweit, Dr. David A. Tyner, Dr. Hans A. Wagner, Gail M. Webber, and Peter K. Yonan.

My colleagues have also brought other spectra of interest to my attention, pointed out literature references, and helped me to evaluate the relative importance of problems which face the organic chemist who wishes to derive the most information from his NMR spectra.

The permission of Varian Associates to reproduce a number of spectra from the Varian Catalogs is appreciated.

I am especially indebted to Dr. Robert T. Dillon who heads the Searle analytical department, to Aristides John Damascus who supervises the spectral laboratory, and to Miss Diana Ede who determined most of the spectra in this book. The support and talent of these people have been essential to this project.

I appreciate permission given to me by G. D. Searle & Company and by my supervisors, Dr. Robert R. Burtner, Dr. Byron Riegel, and Dr. Albert L. Raymond, to undertake this project.

This book would probably have never been completed without the assistance of my wife, Harriett. Although she had already previously experienced the difficulties associated with projects of this type, she enthusiastically not only served as my chief advisor, but also typed the many drafts of the manuscript.

Edward A. Brown and Dr. Robert T. Dillon of G. D. Searle & Company and LeRoy F. Johnson of Varian Associates critically reviewed the entire manuscript. Their suggestions led to many improvements in the accuracy, presentation, and clarity of the text.

Any comments or criticisms concerning this book will be appreciated. Suggested improvements for possible future editions will be especially helpful.

Roy H. Bible, Jr.

# CONTENTS

Key spectra will be found on the following pages:

| | | |
|---|---|---|
| 6 | 94 | 150 |
| 16 | 98 | 166 |
| 34, 36 | 100 | 174 |
| 42 | 102, 104, 106 | 182, 184 |
| 72 | 112 | 188, 190, 192, 194 |
| 86 | 126 | 196, 198 |
| 88 | 134 | 200 |

# INTRODUCTION

Except as noted, all of the spectra were determined on a Varian A-60 instrument at 60 Mcps using 5 — 15%(weight/volume) solutions in $CDCl_3$ with tetramethylsilane as an internal (dissolved in the solution) reference.

Every effort has been made to remove all unnecessary obstacles from the path of the student. The integration measurements, critical numerical peak positions, and molecular formulas are given. Graph paper printed with all three field scales has been employed for most of the spectra. The necessary reference tables of chemical shifts and coupling constants are given in the appendices.

The numbering of bands is the same for all spectra obtained for the same compound under different conditions. Bands assigned to specific protons have been placed in parentheses if the band also contains signals due to other protons.

The integrations are expressed in millimeters as measured on the original graphs. Division of these values by 5 will give values measured in the units seen on the grid. Thus, a given value of 10 corresponds to 2 units on the grid. Where necessary, short vertical lines are used to indicate the limits used for the measurement of the integration.

The values given for the peak positions and peak separations are those measured on the original spectra. Measurements for the reproductions which appear in this volume cannot easily be made to this same precision.

It is recommended that at 60 Mcps both the 1000 to 500 cps (16.67 to 8.33 ppm) region and the 500 to 0 cps (8.33 to 0 ppm) region be examined routinely. However, to help simplify the presentations in this book, the scan from 1000 to 500 cps has been omitted in many of the spectra.

The exercises have, in general, been arranged in order of increasing difficulty. At selected points, however, this sequence of increasing difficulty has been interrupted by the introduction of easier problems. An attempt to introduce variety has been made through the various approaches taken in the questions.

Wherever possible, principles are mentioned in connection with other problems before these principles are discussed in detail. Once introduced, the fundamentals, particularly those which usually cause trouble, have been emphasized frequently in the remaining discussions.

Coupling constants are treated as part of the problem of spin-pattern analysis. Specific problems involving long-range couplings have been placed in Section 9. Higher-order effects, including "virtual" coupling, are introduced in Section 6.

Even the casual user of NMR spectra soon learns that often a tremendous amount of information can be obtained from a single spectrum. The limitation on the amount of data which is extracted is frequently established by the time which the student wishes to spend in analyzing the spectrum. The discussions of most of the spectra in this book have been purposely kept brief.

Part of the usefulness of this workbook will depend on the ease with which particular examples can be located. For this reason, the general concepts covered in each section are listed in the table of contents, while specific topics and principles are given in the index. A relatively compact molecular-formula index, which also shows the structures of the compounds, permits the student to quickly glance through all of the molecular structures.

References for further study have been made to specific sections in the pertinent textbooks. These texts, which for convenience are referred to by the names of the authors, are as follows:

1. Bhacca, N.S., and D.H. Williams, Applications of NMR Spectroscopy in Organic Chemistry (Illustrations from the Steroid Field) Holden-Day, Inc. (San Francisco), 1964.

2. Bible, R.H., Jr., Interpretation of NMR Spectra: An Empirical Approach, Plenum Press (New York), 1965.
3. Emsley, J.M., J. Feeney, and L.H. Sutcliffe, High Resolution Nuclear Magnetic Resonance Spectroscopy (in two volumes), Pergamon Press (Oxford), 1965 and 1966.
4. Jackman, L.M., Applications of Nuclear Magnetic Resonance Spectroscopy in Organic Chemistry, Pergamon Press (New York), 1959.
5. Pople, J.A., W.G. Schneider, and H.J. Bernstein, High-Resolution Nuclear Magnetic Resonance, McGraw-Hill (New York), 1959.
6. Roberts, J.D., Nuclear Magnetic Resonance, McGraw-Hill (New York), 1959.

A review of this workbook can be made by a study of the "key spectra," which are listed on page xi. The discussions of these spectra summarize the points made in earlier spectra and ensure that the necessary principles have been well-established in the reader's mind.

# SECTION 1

General considerations, appearance of a good spectrum, concept of the chemical shift, use of the integration curve, and $D_2O$ exchange.

*Suggested Reading*

Bhacca and Williams: pp. 1-9 and 11.
Bible: pp. 1-30.
Emsley, Feeney, and Sutcliffe: pp. 1-9, 59-65, 140-151, 256-274, 280-287, 310-311, 665-726, 749-794, and 838-841.
Jackman: pp. 1-20, 35-49, and 105.
Pople, Schneider, and Bernstein: pp. 1-9 and 87-91.
Roberts: pp. 1-33.

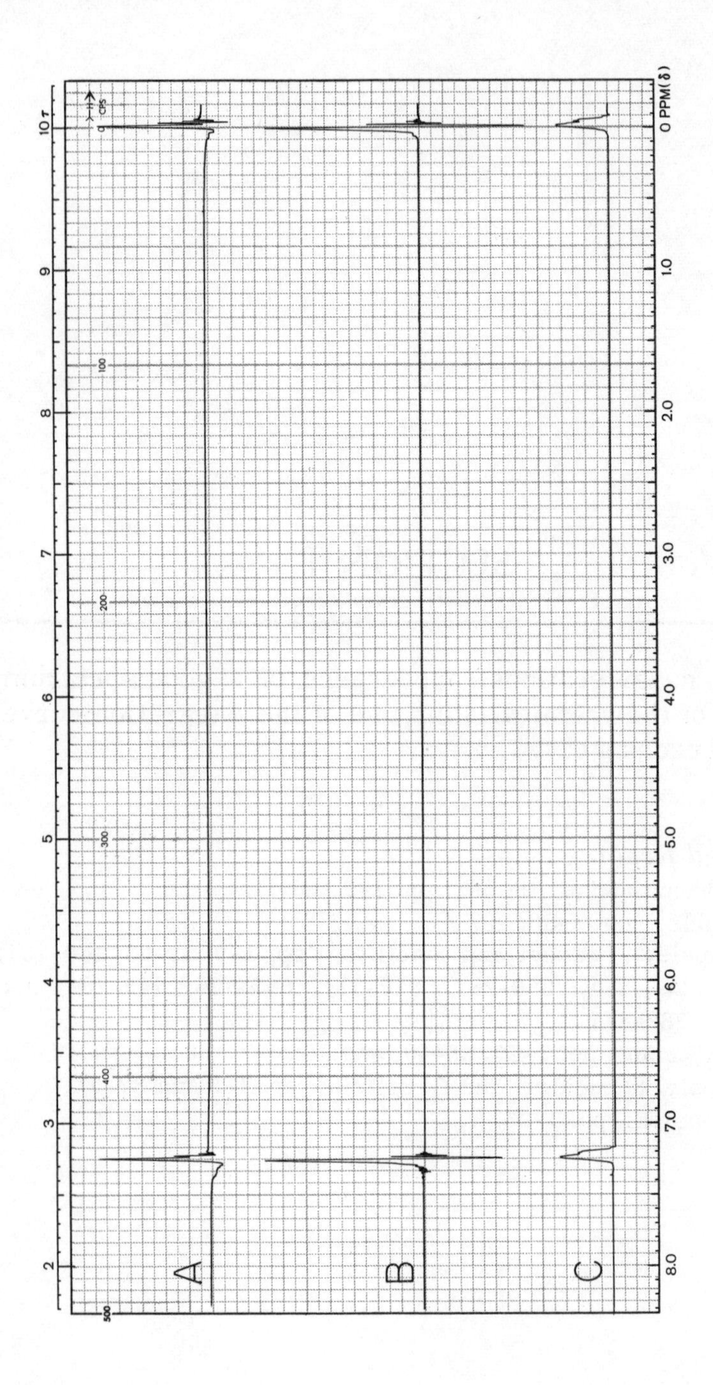
A
B
C
0 PPM(δ)
1.0
2.0
3.0
4.0
5.0
6.0
7.0
8.0
10τ
9
8
7
6
5
4
3
2
100
200
300
400
500

All three of these scans were determined using a mixture of tetramethylsilane [$(CH_3)_4Si$ or TMS] and chloroform ($CHCl_3$). The peak on the left (at lower magnetic field) is due to the proton in the chloroform.

---

*Questions*

1. Which of the three scans (A, B, or C) was determined under the best operating conditions?
2. What is the principal reason for the chloroform peak being at such low field?
3. What is the separation of the two peaks called?
4. Express the separation of the two peaks in each of the three scales (cps, tau, and delta units).
5. What would the separation of the peaks be at 100 Mcps? Give the answer in each of the three scales.

Answers: p. 203.

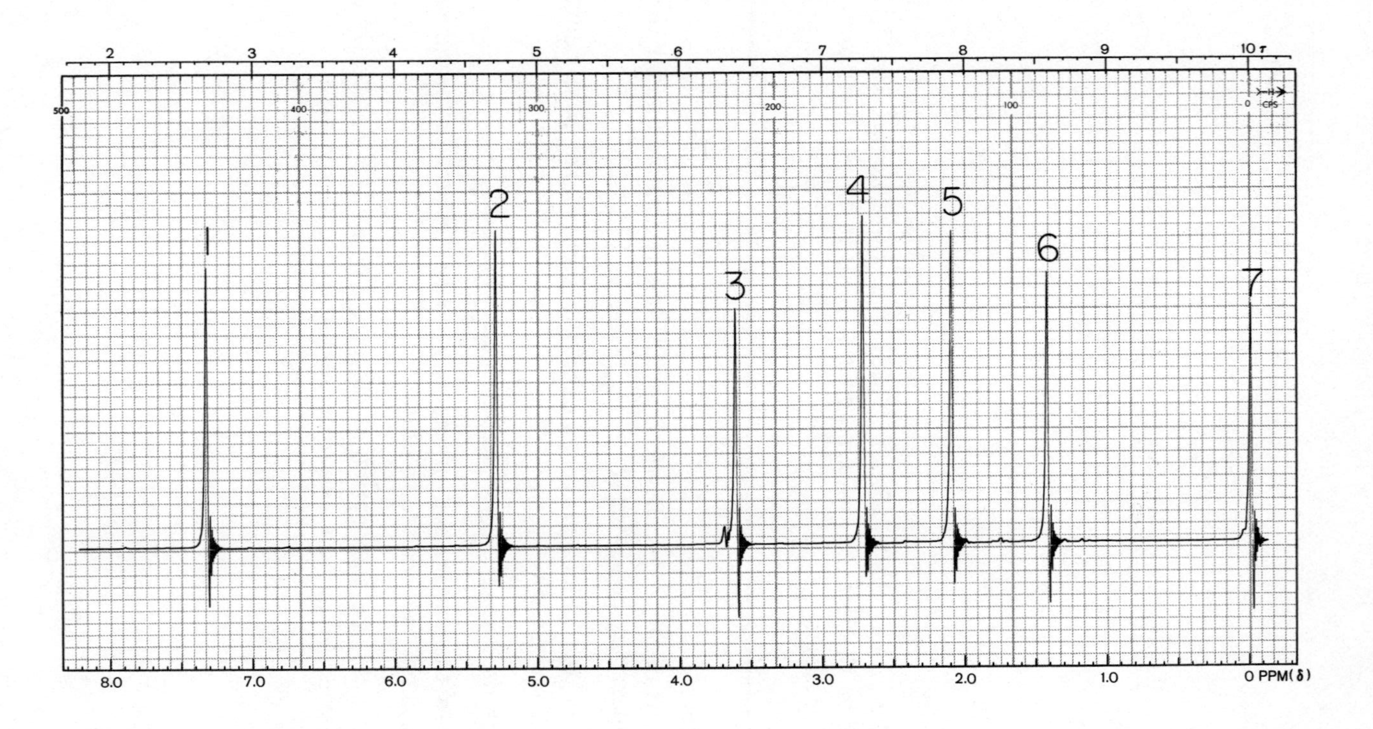

2
3
4
5
6
7
8
9
10 τ
500
400
300
200
100
0 CPS
1
2
3
4
5
6
7
8.0
7.0
6.0
5.0
4.0
3.0
2.0
1.0
0 PPM(δ)

This spectrum was determined using a solution of the following components in deuterochloroform ($CDCl_3$):

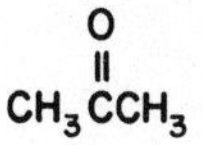

$CH_3COCH_3$ (I), $CHCl_3$ (II), cyclohexane $(CH_2)_6$ (III), $CH_2Cl_2$ (IV), 1,4-dioxane $O(CH_2CH_2)_2O$ (V), $(CH_3)_4Si$ (VI), $Cl_3CCH_3$ (VII)

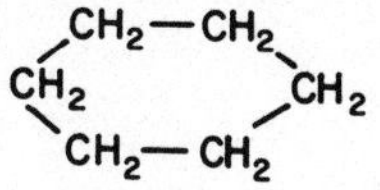

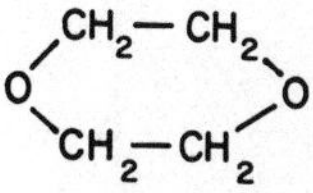

This mixture has been recommended for checking the calibration of the Varian A-60 instrument. [J. L. Jungnickel, Anal. Chem. 35:1985 (1963).]

---

*Questions*

1. Match each of the components with the corresponding peak (⑦ equals VI, etc.).
2. What would be the position (cps) of the chloroform peak at 60 Mcps with reference to internal cyclohexane?

Answers: p. 204.

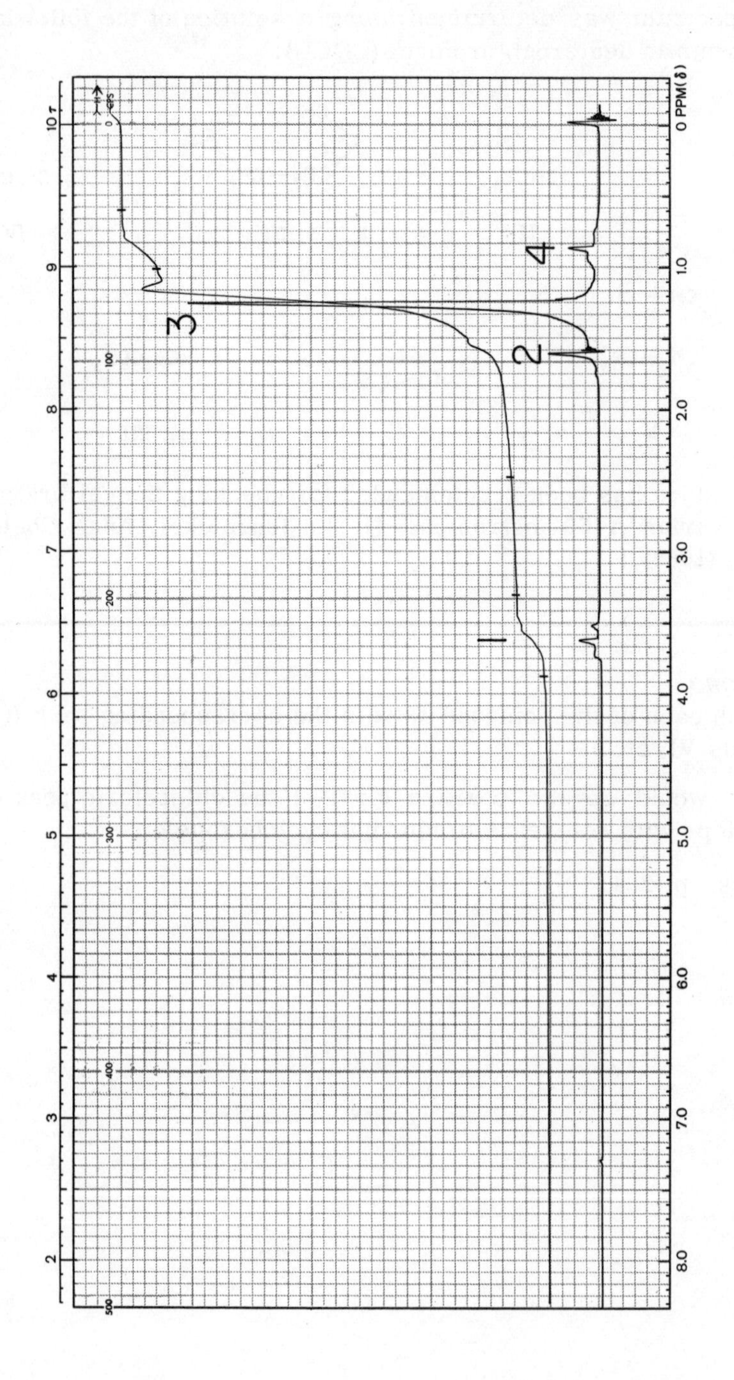

10τ
9
8
7
6
5
4
3
2
0 PPM(δ)
1.0
2.0
3.0
4.0
5.0
6.0
7.0
8.0
0
100
200
300
400
500
3
2
1
4

This is the spectrum of an aliphatic alcohol of the type $CH_3(CH_2)_xCH_2OH$. The spectrum after $D_2O$ exchange is shown on p. 12.

| Band | ① | ② + ③ | ④ | Sum |
|---|---|---|---|---|
| Integration | 11.1 | 149 | 15 | 175.1 |

---

*Questions*

1. Identify the alcoholic proton signal.
2. Assign the bands due to the $C\underline{H}_3$, $C\underline{H}_2OH$, and the remaining $C\underline{H}_2$ protons.
3. Estimate the value of $x$ by using the integration curve.

Answers: p. 205.

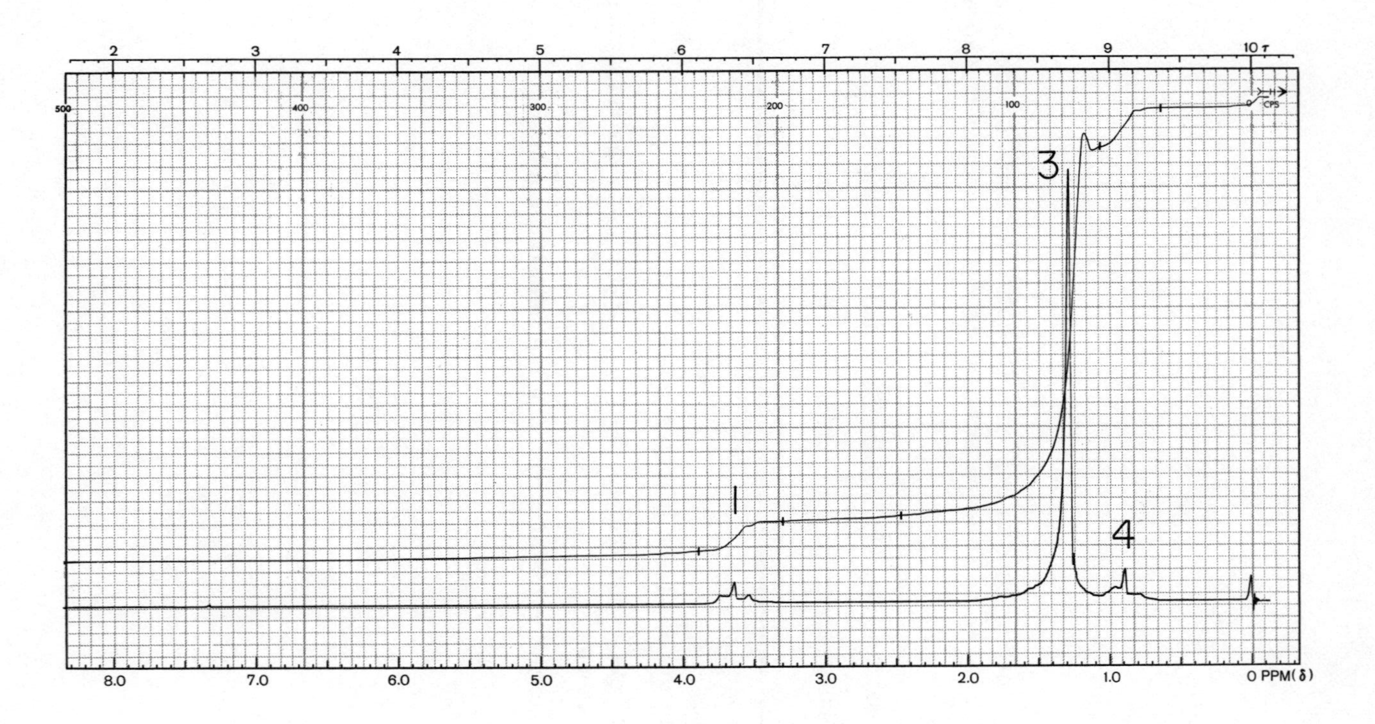
2
3
4
5
6
7
8
9
10 τ
500
400
300
200
100
0
CPS
3
1
4
8.0
7.0
6.0
5.0
4.0
3.0
2.0
1.0
0 PPM(δ)

This is the spectrum determined after $D_2O$ exchange of the solution used for the spectrum on p. 10.

---

*Question*

1. What could be done to increase the accuracy of the integration?

Answer: p. 205.

# SECTION 2

First-order spin systems, presence of impurities, and an unknown.

*Suggested Reading*

Bhacca and Williams: pp. 9-10 and 49-61.
Bible: pp. 30-55.
Emsley, Feeney, and Sutcliffe: pp. 166-183.
Jackman: pp. 20-24 and 83-88.
Pople, Schneider, and Bernstein: pp. 9-10, 91-98, and 192-195.
Roberts: pp. 42-48 and 53-55.

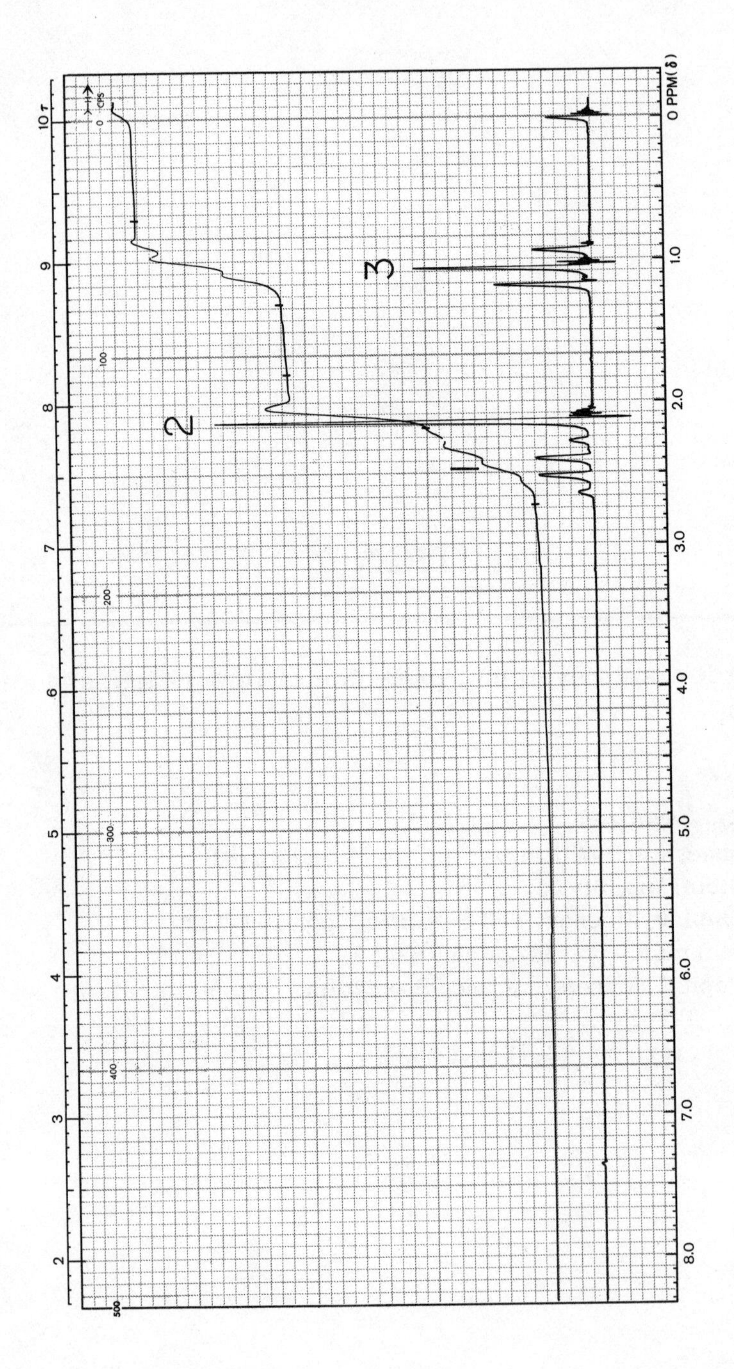

10 τ
9
8
7
6
5
4
3
2
0 CPS
100
200
300
400
500
0 PPM (δ)
1.0
2.0
3.0
4.0
5.0
6.0
7.0
8.0
1
2
3

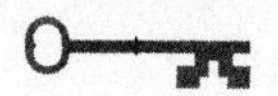

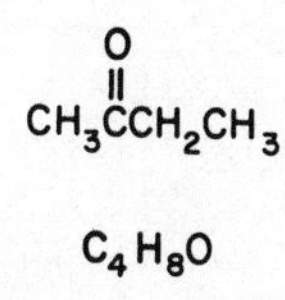


---

*Questions*

1. Assign the numbered bands to specific protons.

2. What two conditions must be met by a system before first-order splitting rules apply?

3. Does this system meet these conditions?

4. Describe the characteristics of first-order multiplets in terms of: (1) the number of peaks, (2) the separation of adjacent peaks, and (3) the relative intensities of the peaks in the multiplets.

5. Draw a first-order splitting diagram for the $-CH_2CH_3$ pattern.

6. The protons in the isolated methyl group are undoubtedly coupled with each other [$J \approx -12.4 + (-1.9) \approx -14.3$ cps]. Why is the signal due to these three protons sharp and unsplit?

Answers: p. 206.

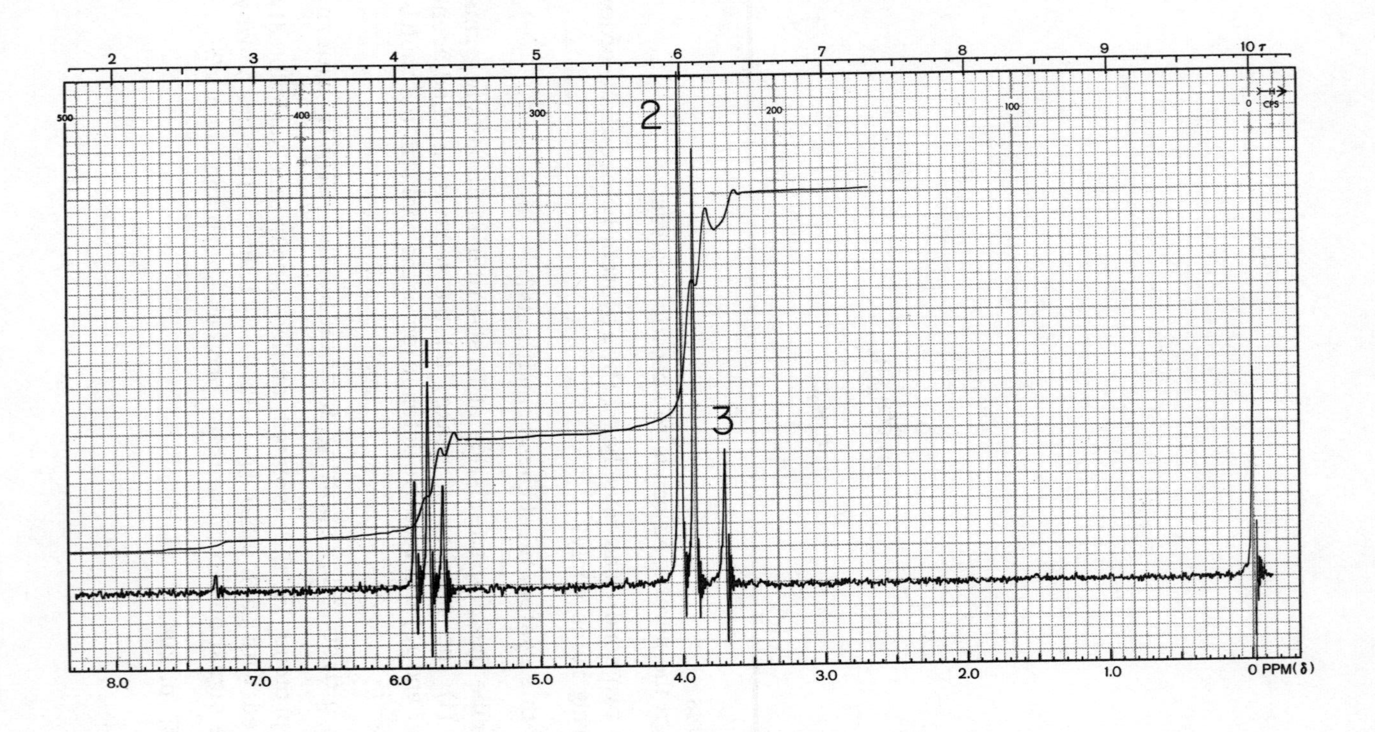

10 τ
9
8
7
6
5
4
3
2
0 CPS
100
200
300
400
500
0 PPM (δ)
1.0
2.0
3.0
4.0
5.0
6.0
7.0
8.0
1
2
3

$CHCl_2CH_2Cl$

$C_2H_3Cl_3$

---

*Questions*

1. Assign the bands to specific protons.
2. Point out two peaks which are obviously due to impurities.
3. What two factors must be considered before it can be assumed that there is equal coupling between the protons in the two groups?
4. Describe this pattern in notational terms.
5. Are the relative intensities of the peaks approximately correct for a first-order pattern?

Answers: p. 207.

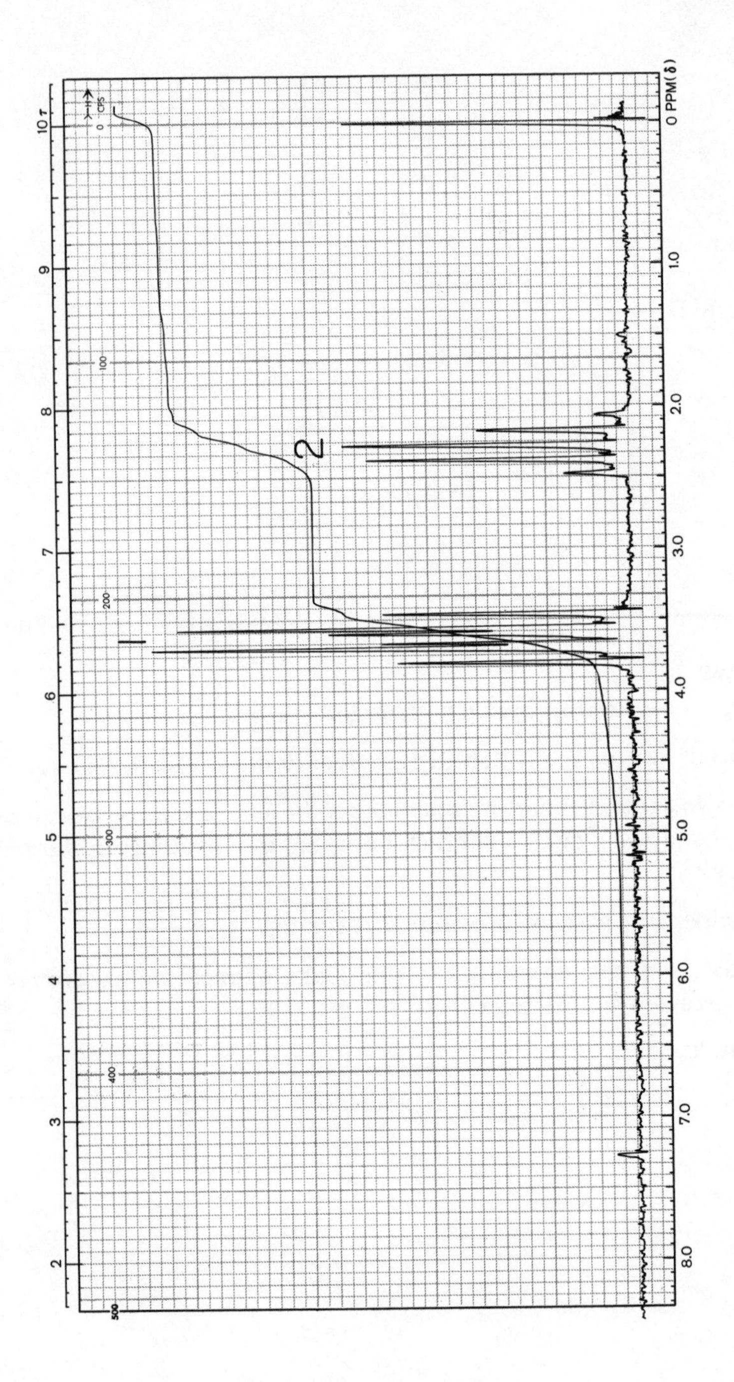

$ClCH_2CH_2CH_2Br$

$C_3H_6BrCl$

---

*Questions*

1. Assign the bands to specific protons.
2. Explain the splittings.
3. Draw a first-order splitting diagram for this system.

Answers: p. 207.

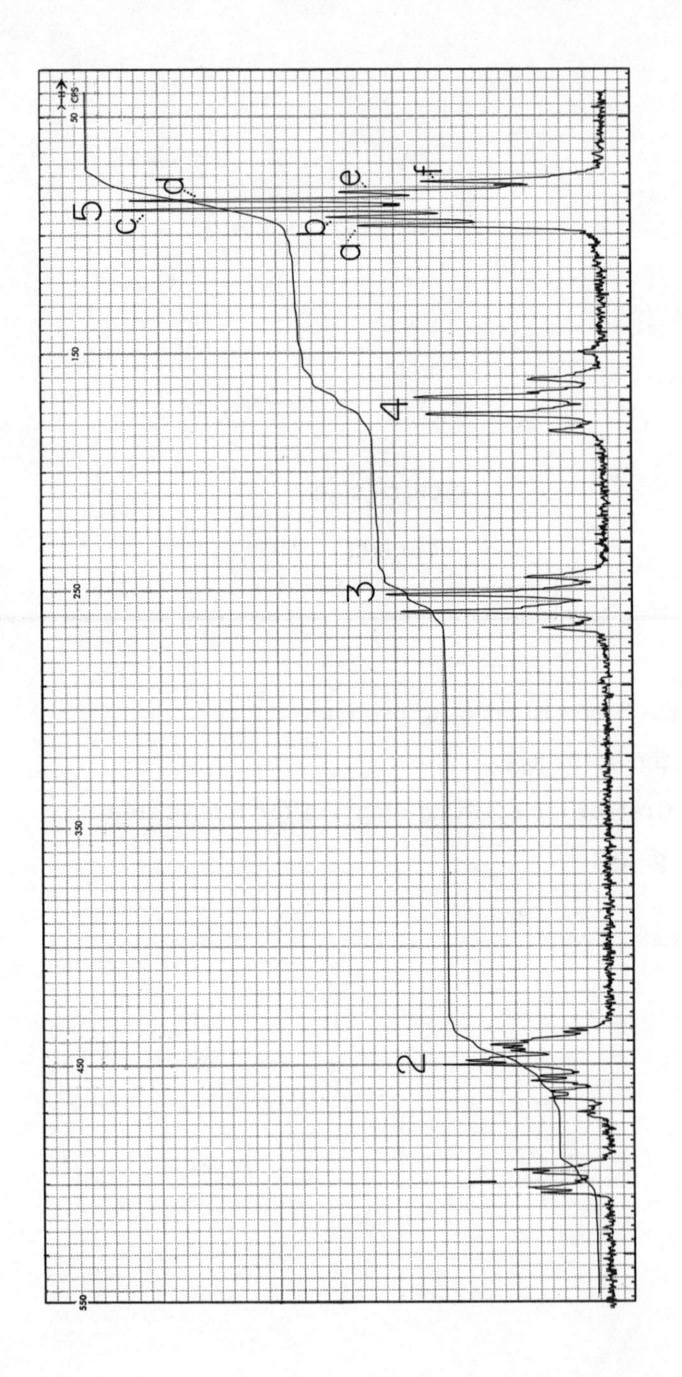

50 CPS
150
250
350
450
550
1
2
3
4
5
a
b
c
d
e
f

$C_{12}H_{14}N_2O$

This spectrum has been offset 50 cps.

| Band | ① | ② | ③ | ④ | ⑤ | Sum |
|---|---|---|---|---|---|---|
| Integration | 15 | 45.5 | 28.1 | 31.5 | 88 | 208.1 |

The small peak at 150 cps (note the offset of the spectrum) is due to an impurity.

---

*Questions*

1. Assign all of the numbered bands to specific protons.
2. Explain the splitting in bands ③, ④, and ⑤.
3. Describe in notational terms the pattern seen in bands ① and ②.

Answers: p. 208.

# SECTION 3

General problems involving chemical shifts, first-order spin systems, exchange rates, and use of the integration curve.

*Suggested Reading*

Bhacca and Williams: pp. 13-41 and 47-49.

Bible: pp. 55-59.

Emsley, Feeney, and Sutcliffe: pp. 481-497, 507-511, 534-551, 726-740, and 816-826.

Jackman: pp. 26-28, 50-74, 103-105, and 112-130.

Pople, Schneider, and Bernstein: pp. 100-102, 218-227, and 417-421.

Roberts: pp. 33-41 and 61-68.

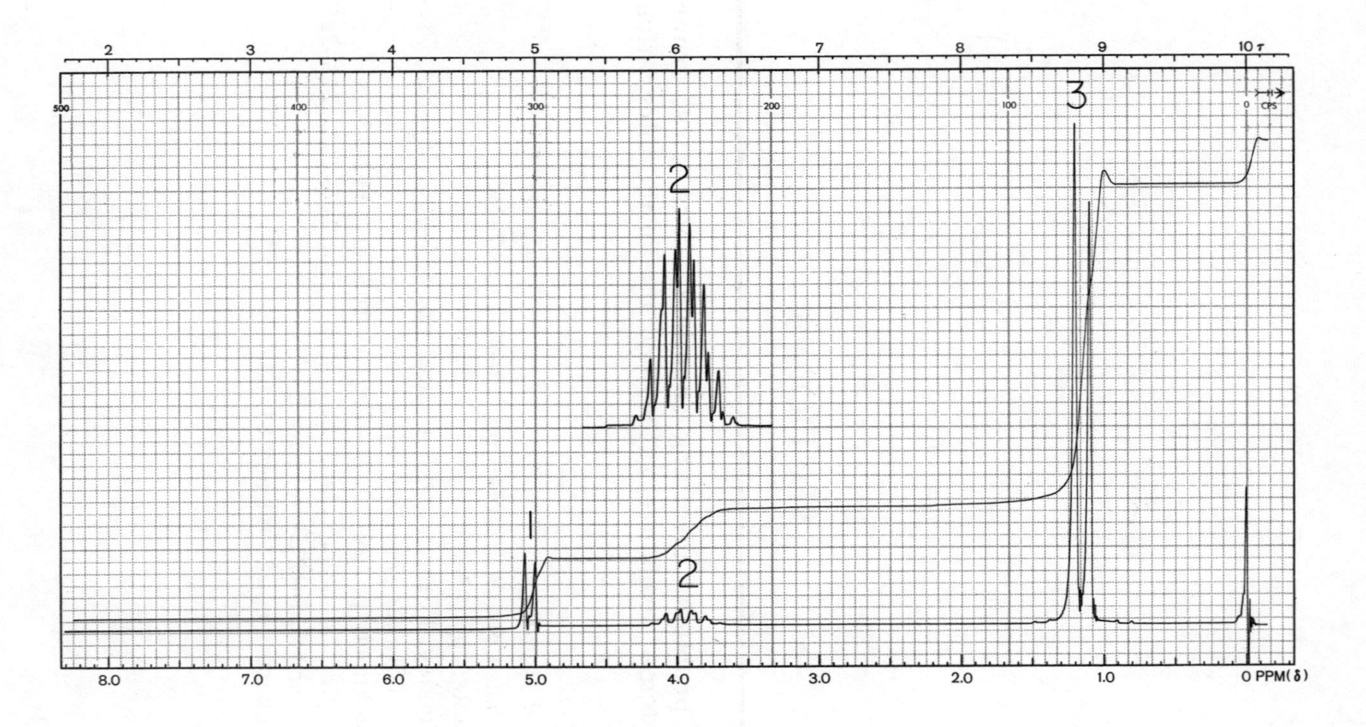
2
3
4
5
6
7
8
9
10 τ
500
400
300
200
100
0
CPS
2
3
1
2
8.0
7.0
6.0
5.0
4.0
3.0
2.0
1.0
0 PPM(δ)

$$
\begin{array}{c}
CH_3 \\
| \\
H-COH \\
| \\
CH_3
\end{array}
$$

$C_3H_8O$

This spectrum was determined using a mixture of isopropanol and tetramethylsilane. No deuterochloroform was added. The 500 to 0-cps region is shown. The portion of the spectrum which is shown in the insert was determined at a higher amplification. The scales of the complete spectrum and of the insert are identical.

| Band | ① | ② | ③ |
|---|---|---|---|
| Integration | 24.5 | 21 | 134 |

---

*Questions*

1. Assign all of the peaks to specific protons.
2. Which of the splittings are first-order?
3. Construct a diagram which will explain all of the multiplicities.

Answers: p. 209.

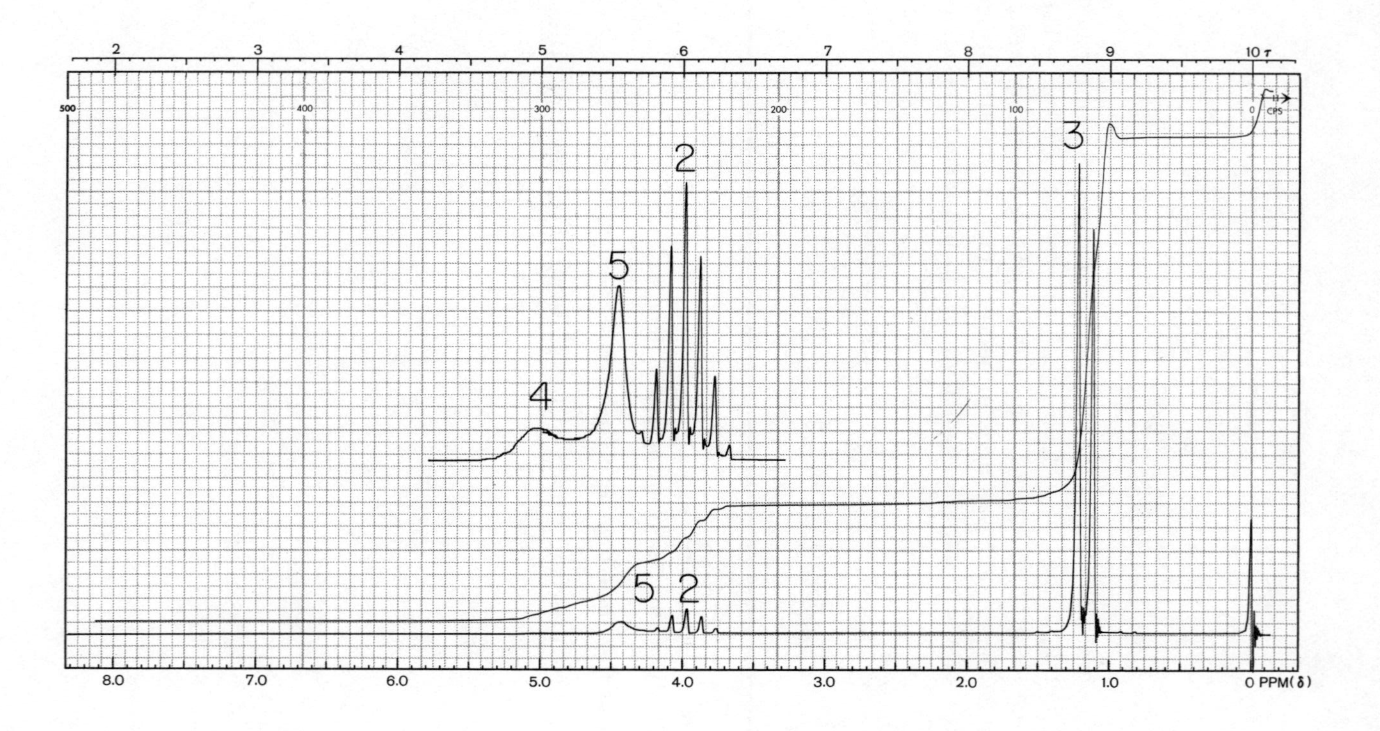
2
3
4
5
6
7
8
9
10 τ
500
400
300
200
100
0
CPS
3
2
5
4
5 2
8.0
7.0
6.0
5.0
4.0
3.0
2.0
1.0
0 PPM(δ)

$$\begin{array}{c} CH_3 \\ | \\ H-COH \\ | \\ CH_3 \end{array} + D_2O$$

The solution used for the spectrum on p. 26 was diluted with a few drops of $D_2O$. The spectrum was then rerun over the region 500 to 0 cps. The portion of the spectrum shown in the insert was obtained by using a higher amplification. The scales of the complete spectrum and of the insert are identical.

---

*Questions*

1. Assign the bands to specific protons.
2. Explain the changes brought about in the spectrum by $D_2O$ dilution.

Answers: p. 210.

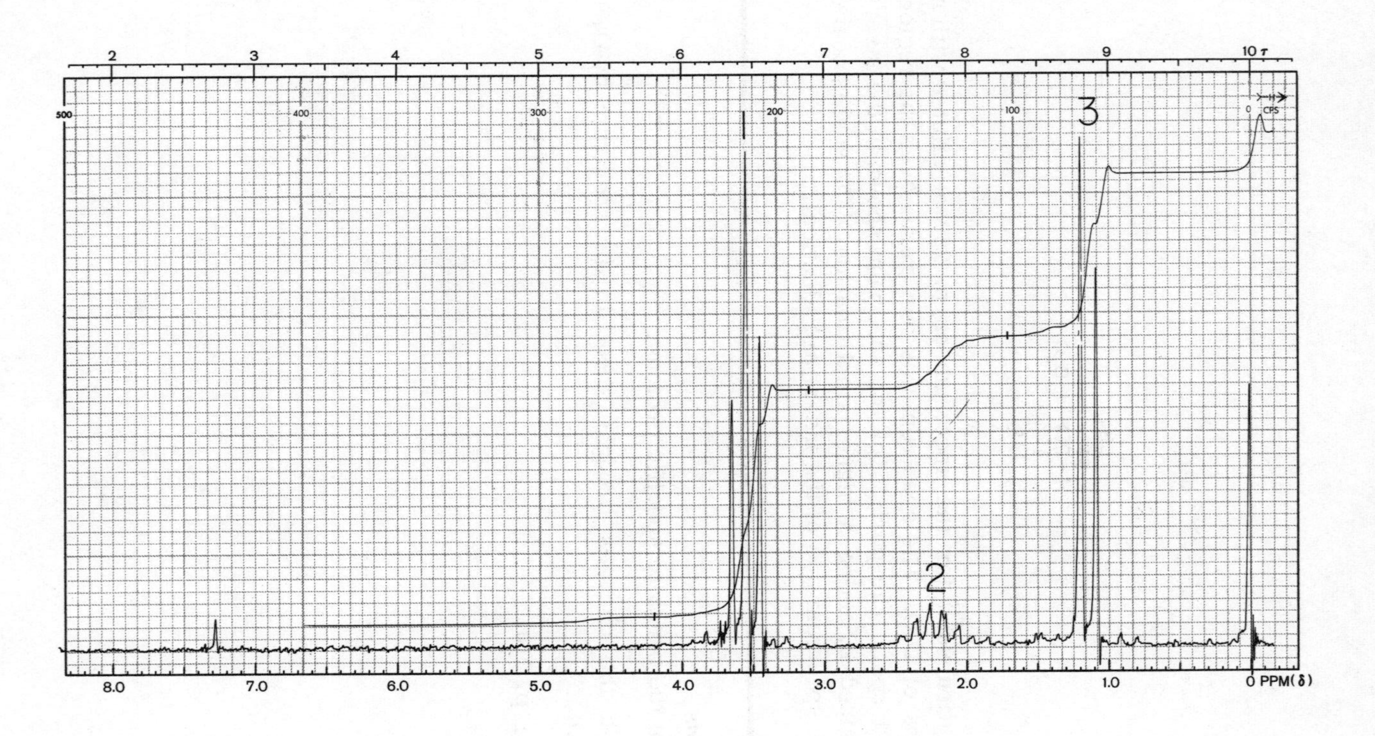
2
3
4
5
6
7
8
9
10 τ
500
400
300
200
100
0 CPS
1
2
3
8.0
7.0
6.0
5.0
4.0
3.0
2.0
1.0
0 PPM (δ)

$$\begin{array}{c} CH_3 \\ | \\ ClCH_2CHCH_2Br \end{array}$$

$C_4H_8BrCl$

| Band | ① | ② | ③ | Sum |
|---|---|---|---|---|
| Integration | 95 | 22 | 65 | 182 |

*Questions*

1. Explain the apparent triplet in band ①.
2. Is there anything in the appearance of the triplet in band ① which is atypical of a first-order multiplet?
3. What experiments might help to confirm the explanation of the triplet?

Answers: p. 210.

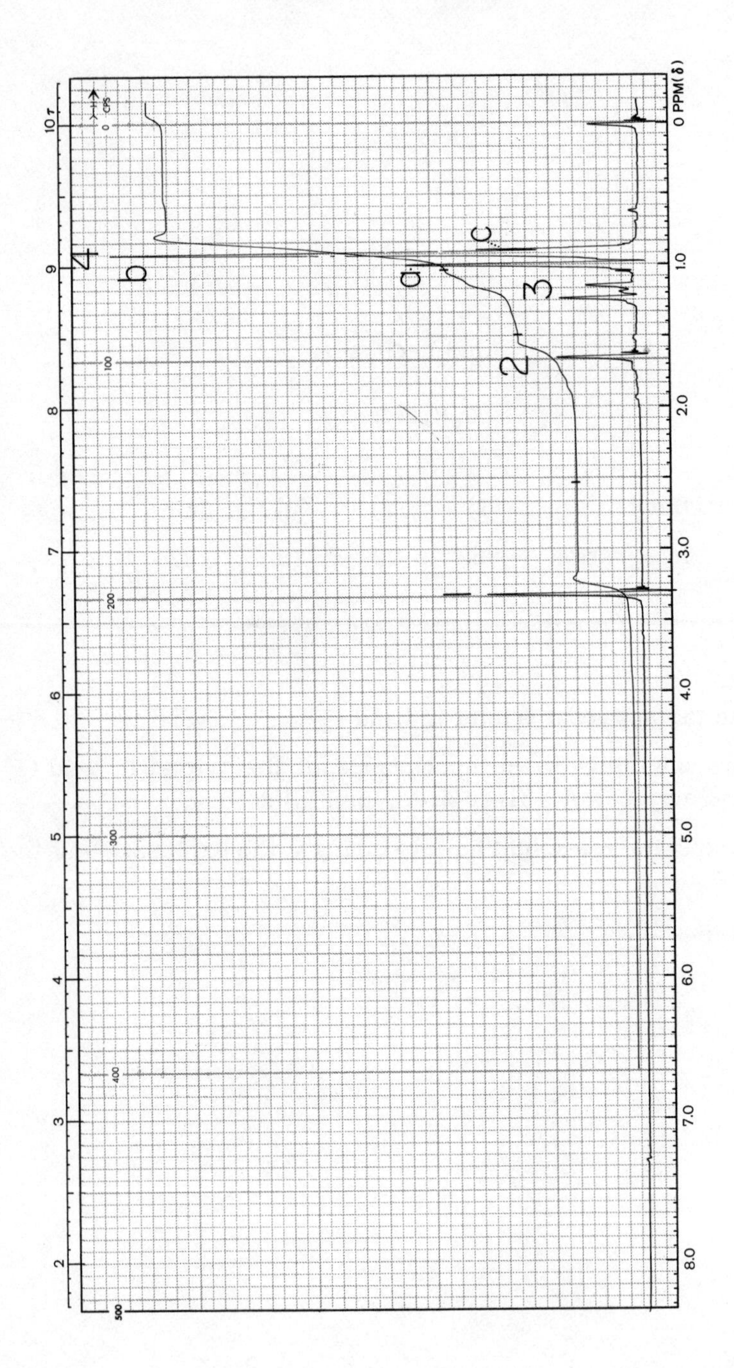

10 τ
9
8
7
6
5
4
3
2
0 PPM (δ)
1.0
2.0
3.0
4.0
5.0
6.0
7.0
8.0
0 CPS
100
200
300
400
500
a
b
c
2
3
4

$$\begin{array}{c} CH_3 \quad\quad\quad\quad CH_3 \\ \diagdown \quad\quad\quad\quad\quad | \\ CH-CH_2-C-CH_2OH \\ \diagup \quad\quad\quad\quad\quad | \\ CH_3 \quad\quad\quad\quad CH_3 \end{array}$$

$C_8H_{18}O$

| Band | ① | ② | ③ | ④ | Sum |
|---|---|---|---|---|---|
| Integration | 23.5 | 23 | 30 | 117.5 | 194 |

The sharp peak in band② is removed by $D_2O$ exchange.

---

*Question*

1. Assign the numbered bands to specific protons.

Answer: p. 211.

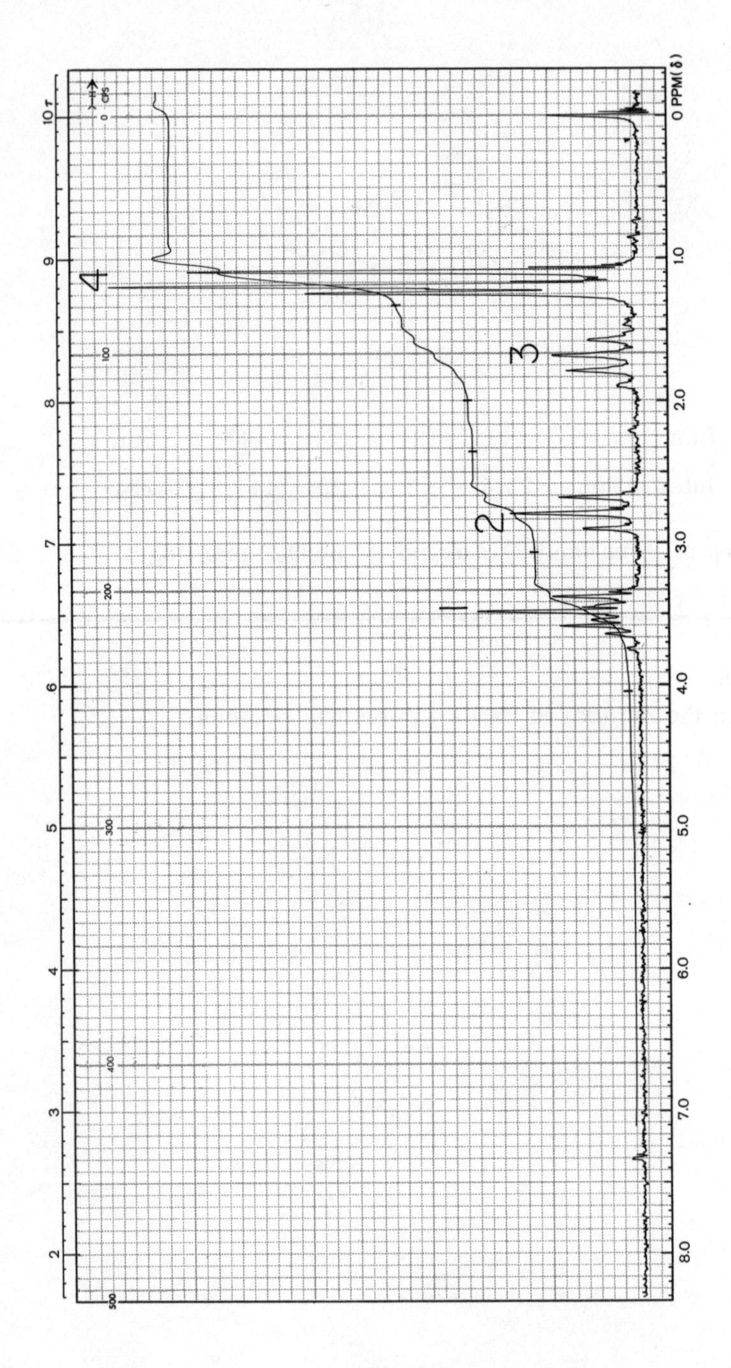

$$\begin{array}{l} CH_3 \\ \quad \diagdown \\ \quad CH-OCH_2CH_2CH_2-NH_2 \\ \quad \diagup \\ CH_3 \end{array}$$

$C_6H_{15}NO$

The spectrum after $D_2O$ exchange is shown on p. 36.

| Band | ① | ② | ③ | ④ | Sum |
|---|---|---|---|---|---|
| Integration | 39 | 25 | 30 | 95 | 189 |

*Questions*

1. Identify the signal or signals due to the $-NH_2$ protons.
2. Explain why most basic amine protons give single sharp signals even though these protons are probably coupled to the nitrogen nucleus and to neighboring protons.
3. Explain all of the observed spin patterns.
4. Are these first-order or higher-order splittings ?

Answers: p. 211.

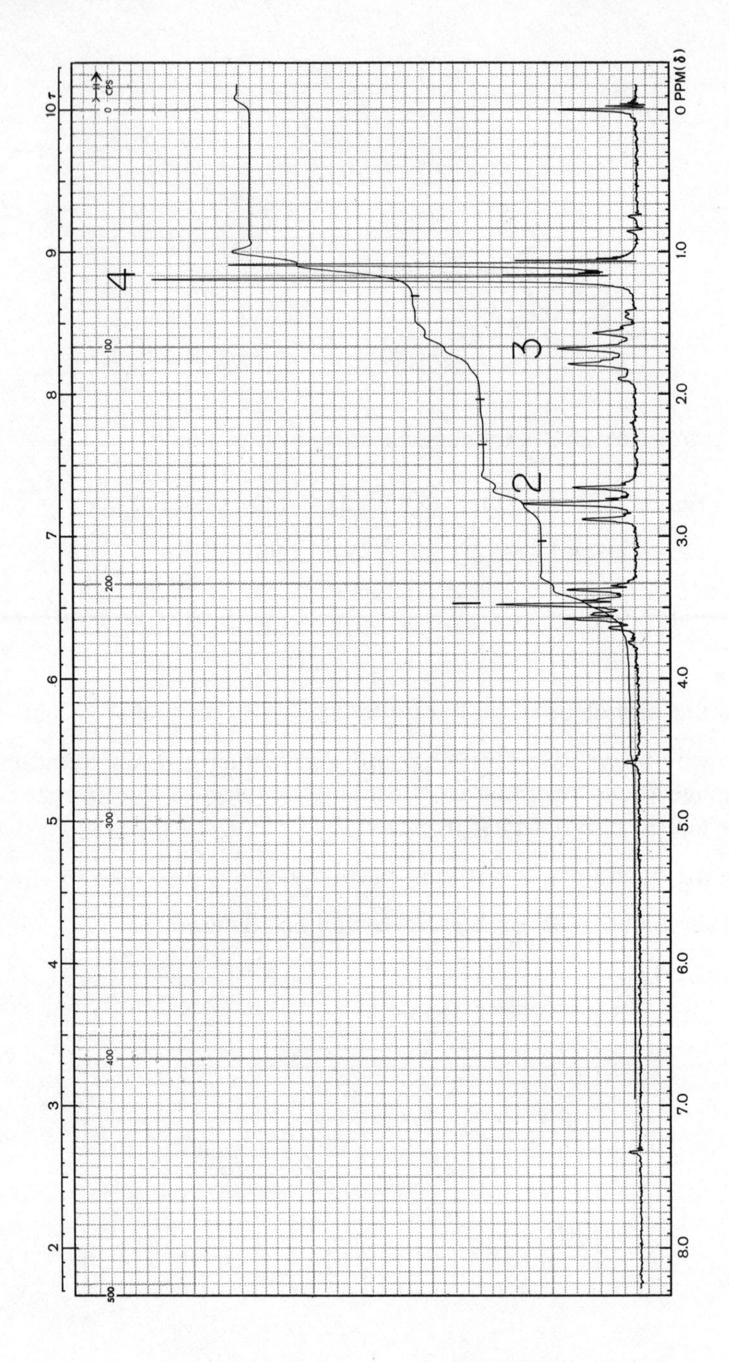

10 τ
9
8
7
6
5
4
3
2
0 CPS
100
200
300
400
500
0 PPM(δ)
1.0
2.0
3.0
4.0
5.0
6.0
7.0
8.0
4
3
2
1

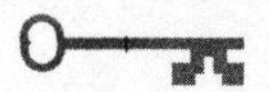

$$\begin{array}{l} CH_3 \\ \quad \diagdown \\ \quad\; CH{-}OCH_2CH_2CH_2{-}ND_2 \\ \quad \diagup \\ CH_3 \end{array}$$

$C_6H_{13}D_2NO$

The spectrum before $D_2O$ exchange is shown on p. 34.

| Band | ① | ② | ③ | ④ | Sum |
|---|---|---|---|---|---|
| Integration | 36 | 28.5 | 28 | 69 | 161.5 |

---

*Questions*

1. What higher-order effect can be detected in these multiplets?
2. What would be the multiplicity of band ③ if this group were coupled significantly more strongly with the $-CH_2ND_2$ protons?

Answers: p. 212.

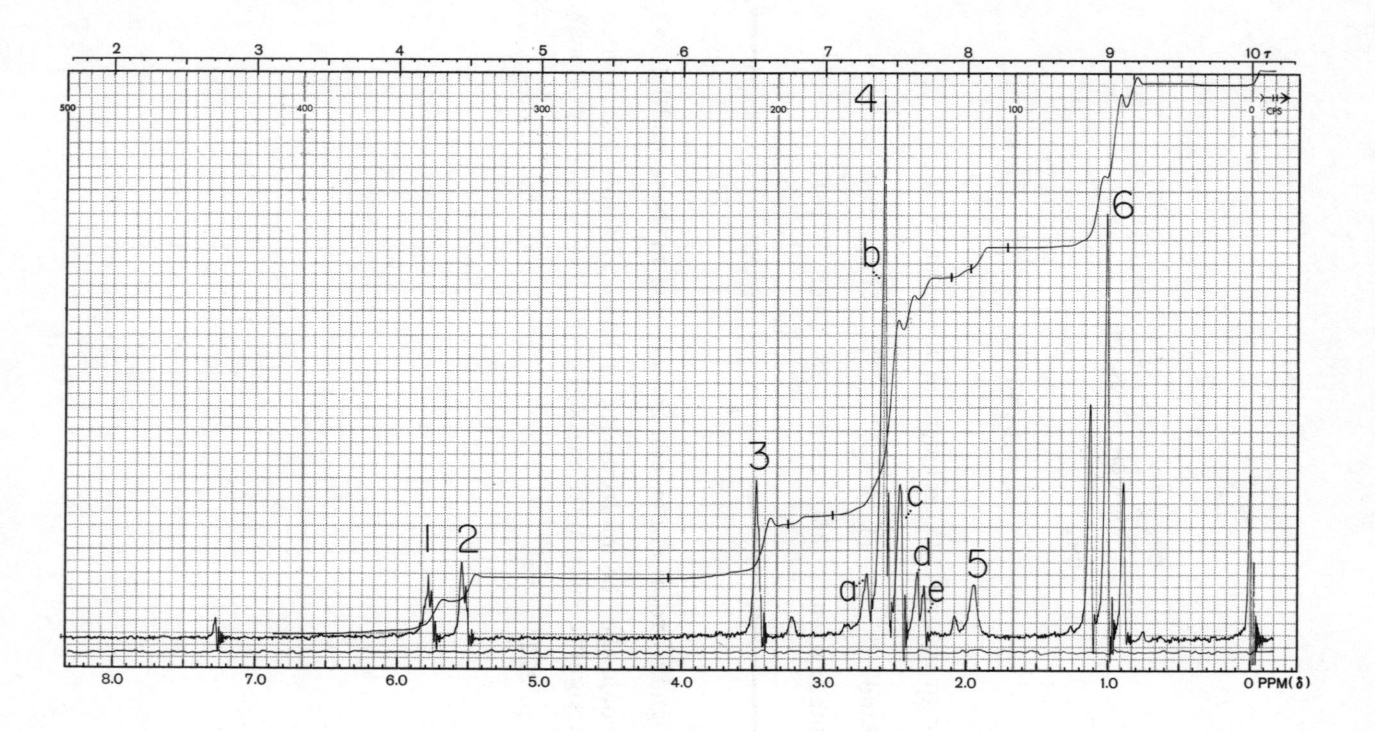

10 τ
CPS
PPM (δ)
1
2
3
4
5
6
a
b
c
d
e

$$\begin{array}{l} H \quad\quad\; Br \\ \;\;\; C{=}C \\ H \quad\quad\; CH_2NHCH_2CH_2N(CH_2CH_3)_2 \end{array}$$

$C_9H_{19}BrN_2$

The lowest trace is the absorption curve over the region 1000 to 500 cps.

| Band | ① | ② | ③ | ④ | ⑤ | ⑥ |
|---|---|---|---|---|---|---|
| Integration | 11.5 | 9.5 | 22 | 99 | 8.5 | 68 |

---

*Questions*

1. Assign all of the numbered bands and lettered peaks.
2. Indicate peaks which are probably due to impurities.

Answers: p. 212.

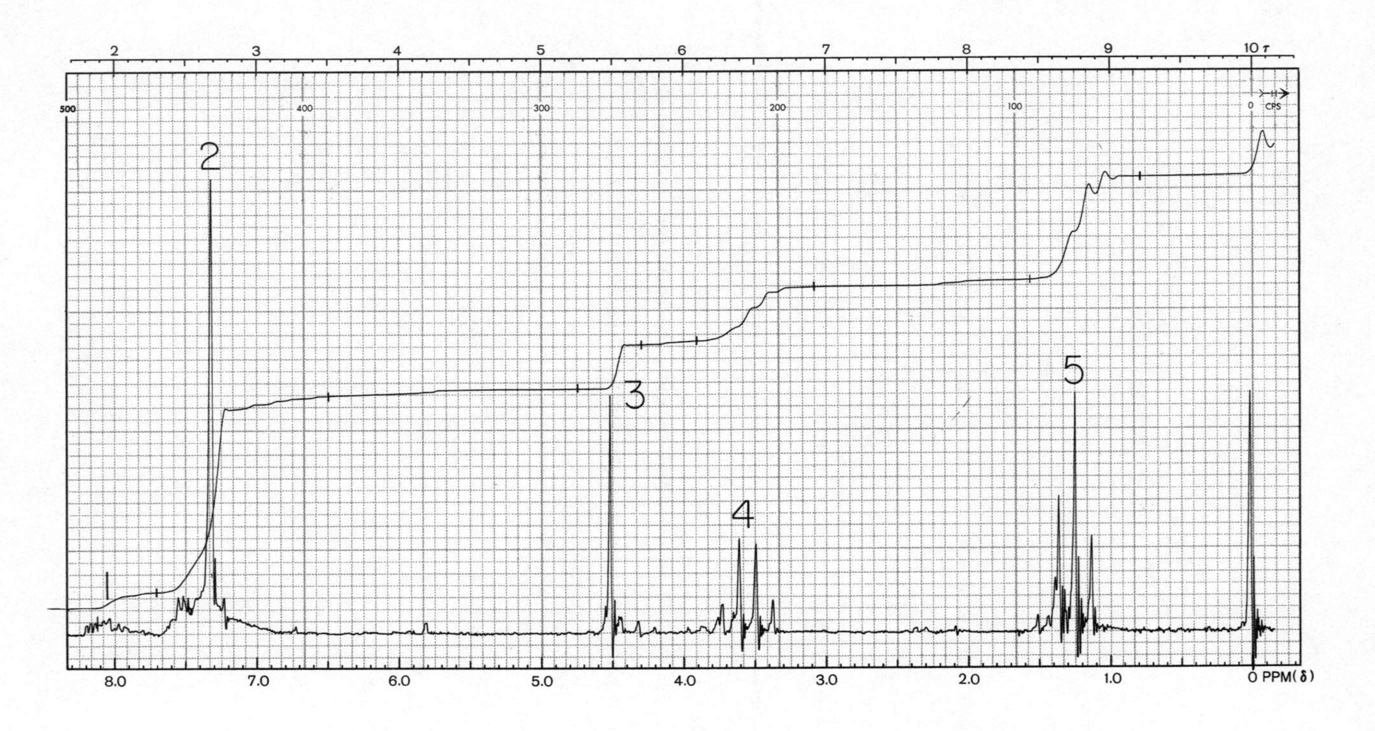
2
3
4
5
6
7
8
9
10 τ
500
400
300
200
100
0
CPS
2
3
4
5
1
8.0
7.0
6.0
5.0
4.0
3.0
2.0
1.0
0 PPM(δ)

This is the spectrum of a sample which was labeled:

$C_6H_5-CH_2OCH_2CH_3$

$C_9H_{12}O$

| Band | ① | ② | ③ | ④ | ⑤ | Sum |
|---|---|---|---|---|---|---|
| Integration | 6.5 | 83 | 19 | 22 | 43 | 173.5 |

---

*Questions*

1. Is this spectrum, including the positions of the methylene proton signals, in general agreement with the proposed structure?
2. Is this sample pure?

Answers: p. 213.

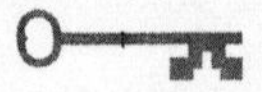

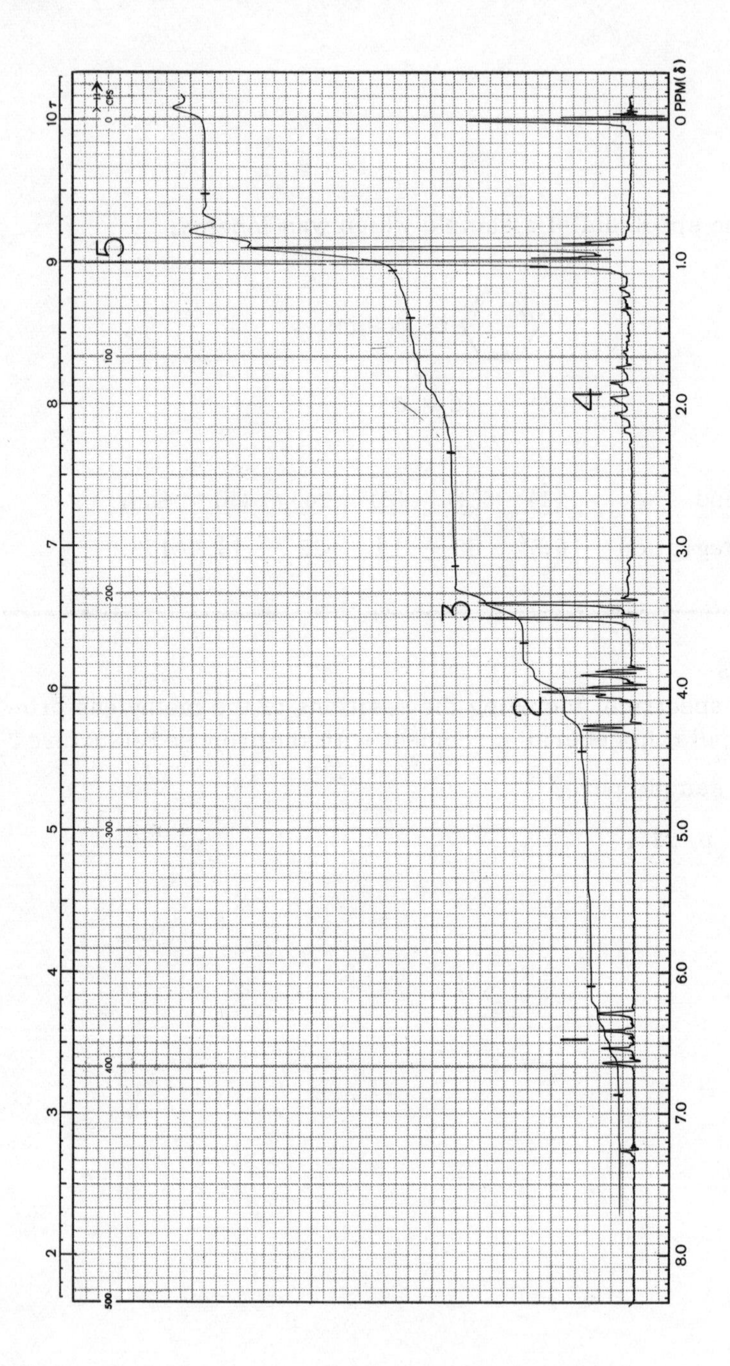

$C_6H_{12}O$

None of the signals are removed by $D_2O$ exchange.

| Band | ① | ② | ③ | ④ | ⑤ | Sum |
|---|---|---|---|---|---|---|
| Integration | 11.5 | 24 | 28 | 17 | 78.5 | 159 |

---

*Questions*

1. Propose a structure for the compound which is consistent with the spectrum.
2. In notational terms, what spin pattern is seen in bands ① and ②?

Answers: p. 213.

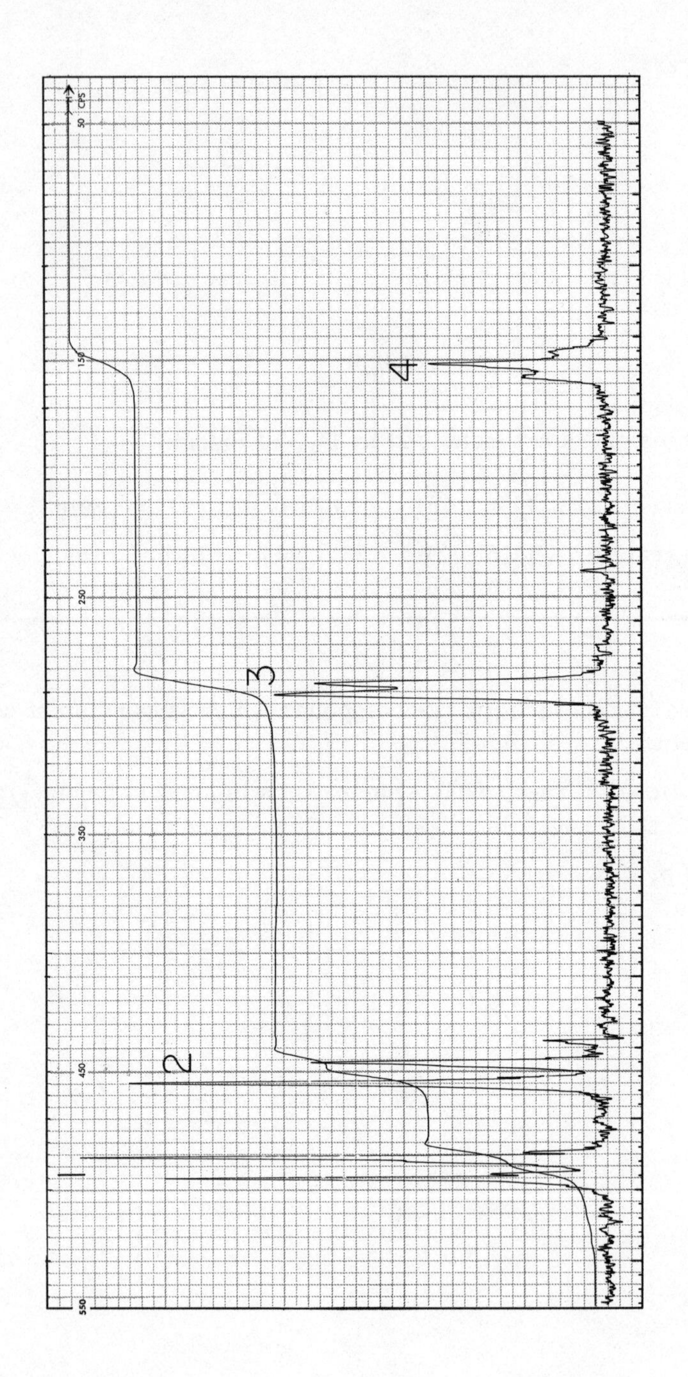
50 CPS
150
250
350
450
550
1
2
3
4

$NO_2-C_6H_4-CH_2OH$

$C_7H_7NO_3$

The spectrum of *p*-nitrobenzyl alcohol was determined on two different occasions. One of the two spectra is given at the left (p. 44), while the other is given on p. 46. On each occasion, a spectrum was also run after $D_2O$ exchange. The spectra after the exchange were identical. One of these spectra is shown on p. 48. Each of the spectra is offset 50 cps.

---

*Questions*

1. Assign the numbered bands.
2. Explain why the two spectra before $D_2O$ exchange are different.
3. Is there anything unusual about the appearance of band ③ ?

Answers: p. 215.

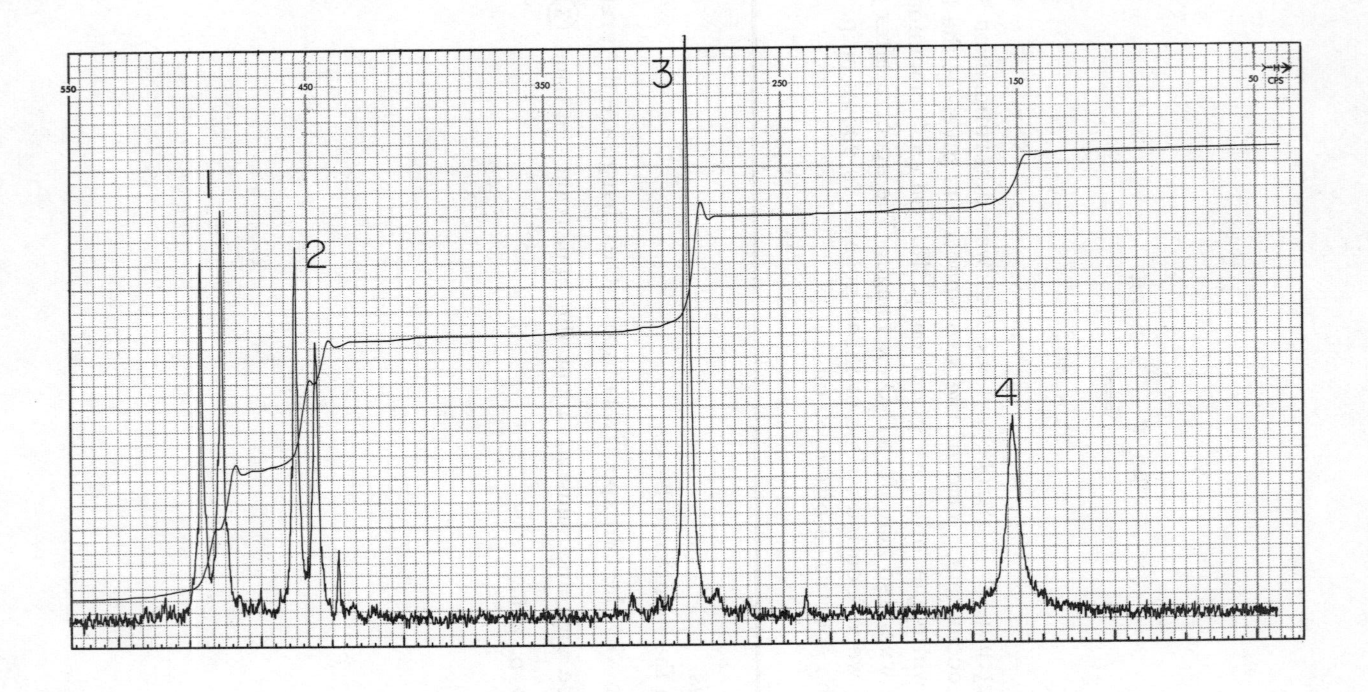
550
450
350
250
150
50
CPS
1
2
3
4

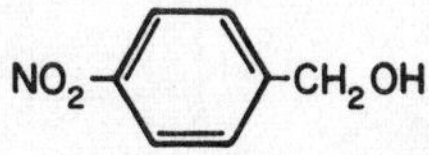

$NO_2-C_6H_4-CH_2OH$

$C_7H_7NO_3$

---

*Questions*

1. What could be done to make bands ③ and ④ sharper?
2. Describe the aromatic protons in notational terms.

Answers: p. 215.

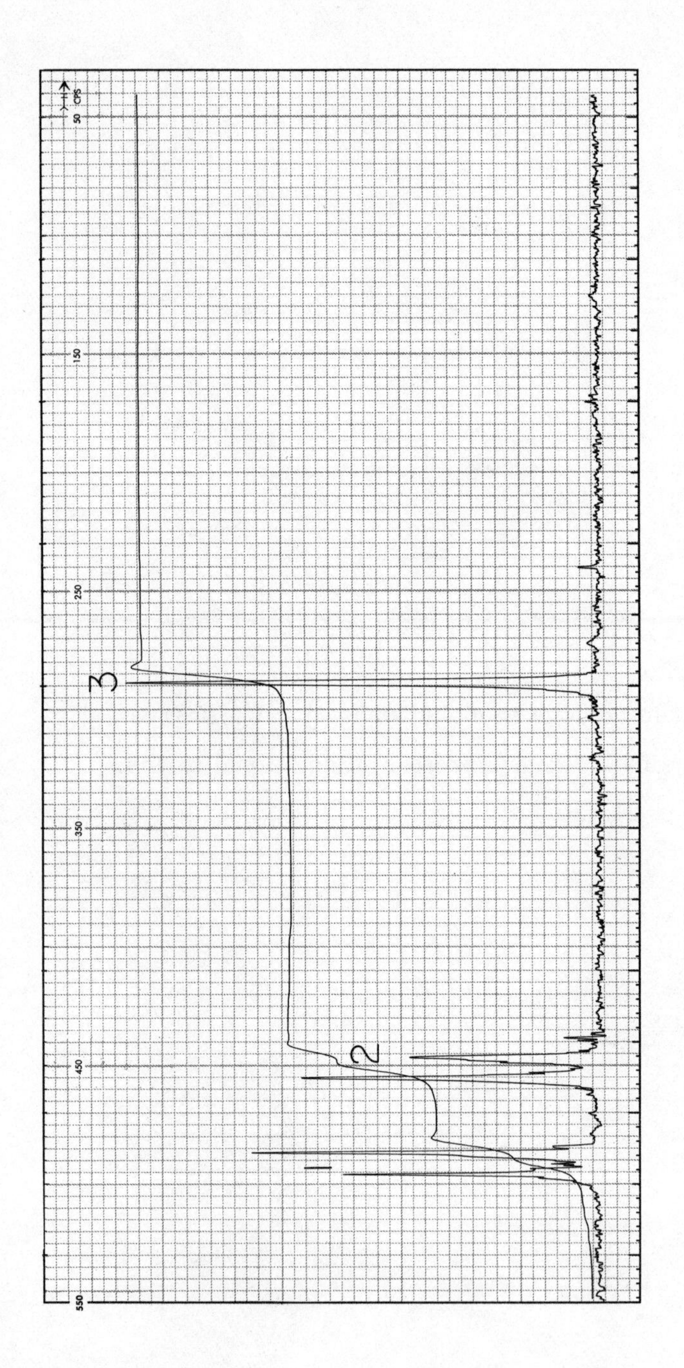

CPS
50
150
250
350
450
550
3
2
1

$NO_2-C_6H_4-CH_2OD$

$C_7H_6DNO_3$

---

*Questions*

1. Suggest an explanation for the slight broadening of the high-field portion (band ②) of the $A'_2B'_2$ pattern. Note that the broadening is also present in the spectra on pp. 44 and 46.

2. How could the explanation for the broadening be verified?

Answers: p. 216.

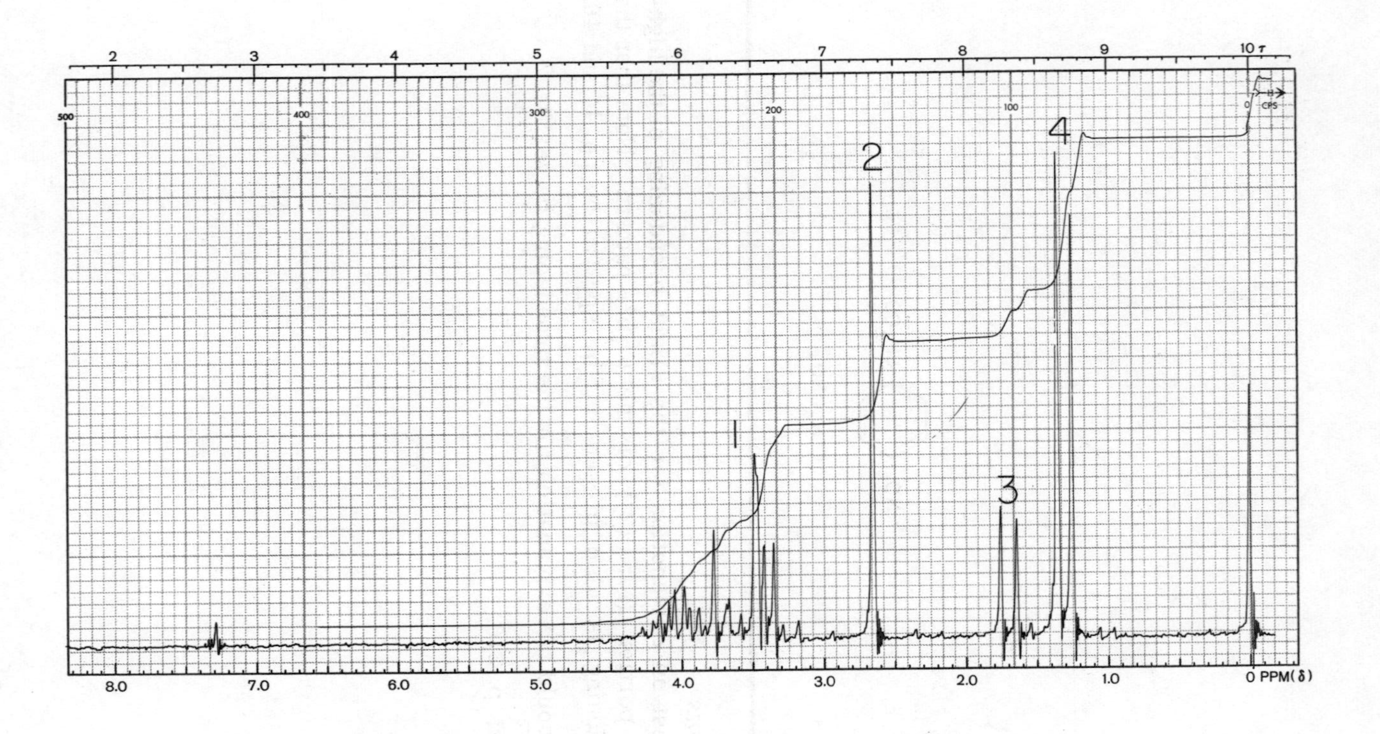

2
3
4
5
6
7
8
9
10 τ
500
400
300
200
100
0
CPS
1
2
3
4
8.0
7.0
6.0
5.0
4.0
3.0
2.0
1.0
0 PPM(δ)

This is the spectrum of a mixture of the following two isomers:

$$\underset{\text{I}}{CH_3-\underset{H}{\overset{Br}{C}}-CH_2OH} \qquad \underset{\text{II}}{CH_3-\underset{H}{\overset{OH}{C}}-CH_2Br}$$

$C_3H_7BrO$

Exchange with $D_2O$ removes band ②, but does not otherwise alter the spectrum. The signal due to the protons of the methyl group in $CH_3CH_2OH$ is at higher field (74 cps) than the signal due to the protons of the methyl group in $CH_3CH_2Br$ (100 cps).

| Band | ① | ② | ③ | ④ | Sum |
|---|---|---|---|---|---|
| Integration | 82.5 | 34.5 | 20.5 | 63 | 200.5 |

---

*Questions*

1. Assign the numbered bands.
2. Calculate the molar percentage of the two isomers in this mixture.

Answers: p. 216.

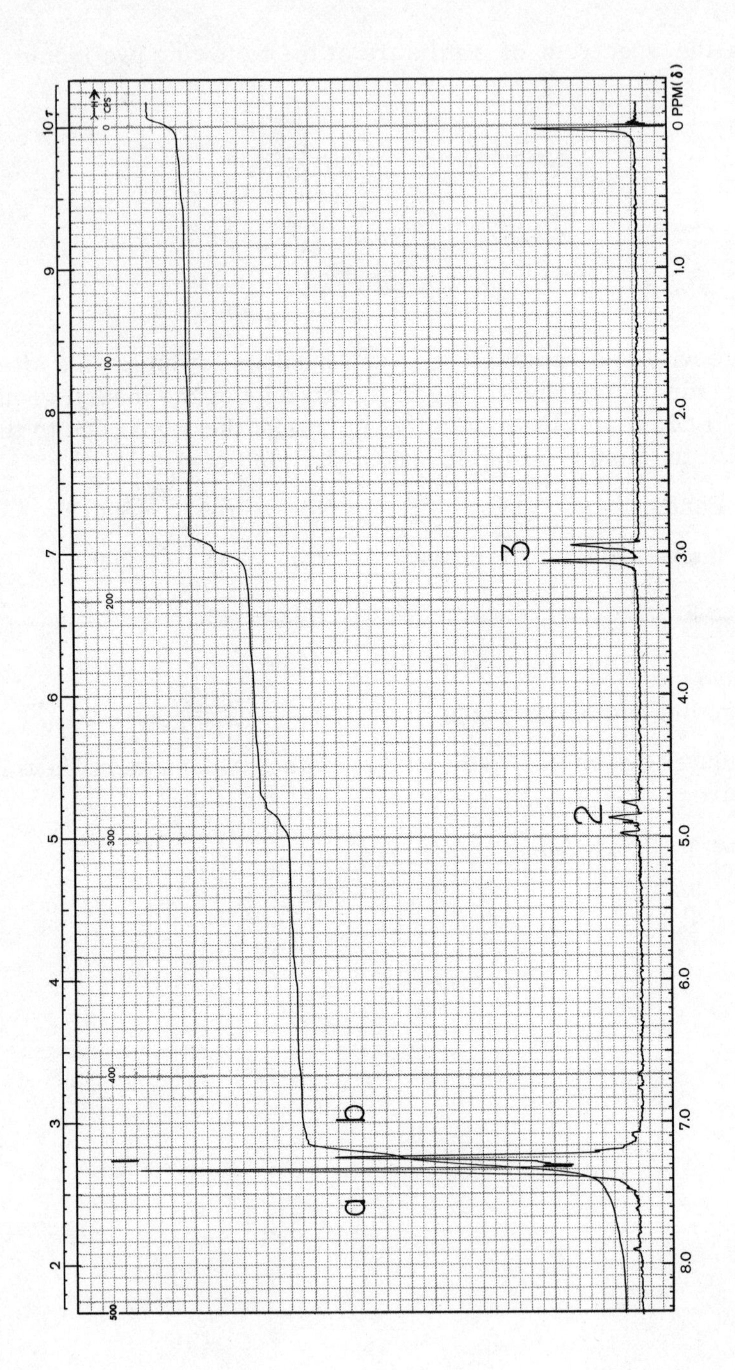
10 τ
9
8
7
6
5
4
3
2
0 PPM(δ)
1.0
2.0
3.0
4.0
5.0
6.0
7.0
8.0
CPS
100
200
300
400
500
a
b
1
2
3

This is the spectrum, after $D_2O$ exchange, of a compound having the molecular formula $C_{14}H_{14}O$. The $D_2O$ exchange removed a sharp signal at 122 cps (2.03 ppm), which was due to one proton.

| Band | ① | ② | ③ | Sum |
|---|---|---|---|---|
| Integration | 134 | 12 | 23 | 169 |

---

*Question*

1. Propose a structure for this compound that is consistent with the spectrum.

Answer: p. 217.

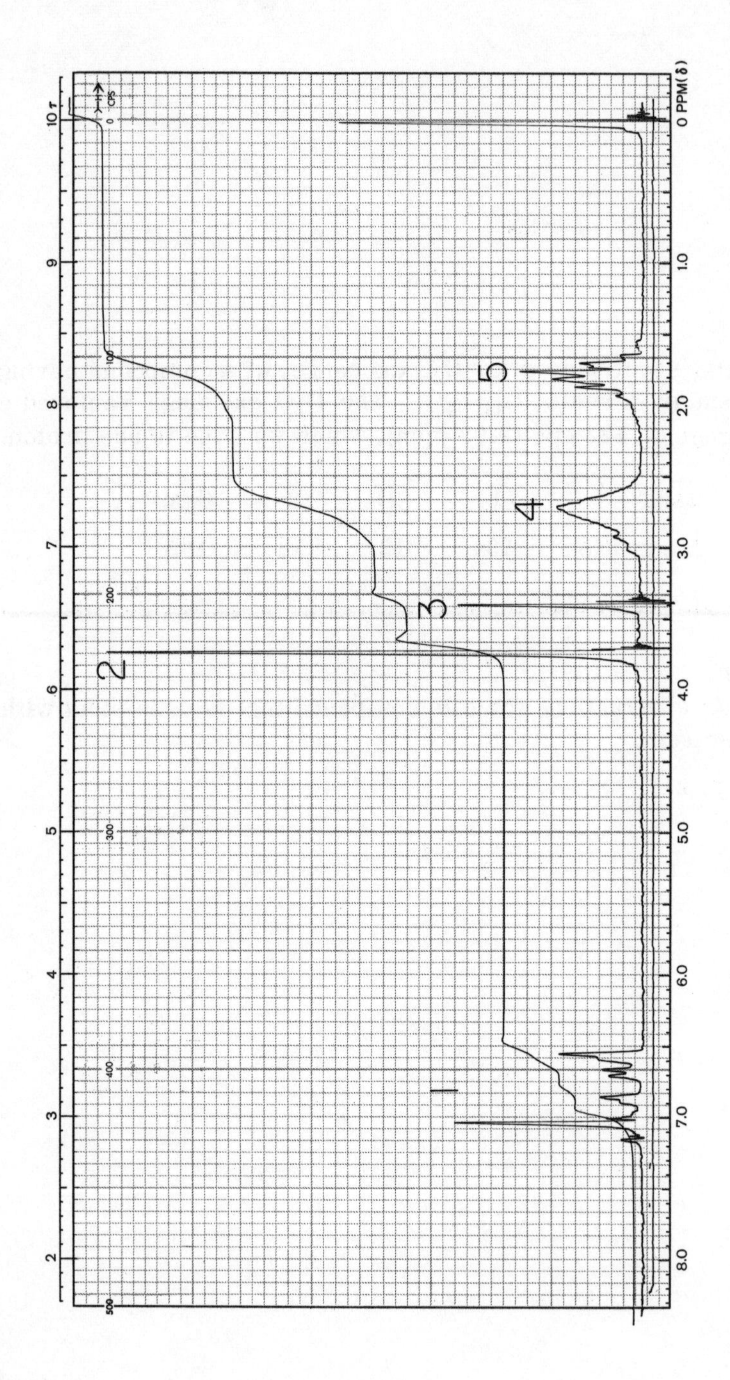

10 τ
9
8
7
6
5
4
3
2
0 PPM (δ)
1.0
2.0
3.0
4.0
5.0
6.0
7.0
8.0
0
100
200
300
400
500
CPS
1
2
3
4
5

This spectrum was obtained using a mixture of the following two isomeric compounds:

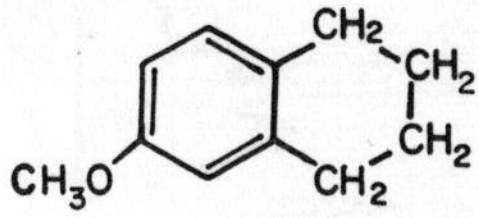

I

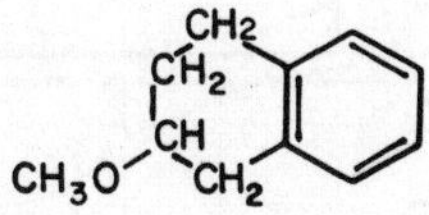

II

$C_{11}H_{14}O$

The lowest trace is the absorption curve from 1000 to 500 cps.

| Band | ① | ② | ③ | ④ | ⑤ | Sum |
|---|---|---|---|---|---|---|
| Integration | 54 | 40.5 | 13.5 | 60 | 54 | 222 |

---

*Questions*

1. Assign the absorption bands to the protons in the two structures.
2. Calculate the molar composition of the mixture.

Answers: p. 217.

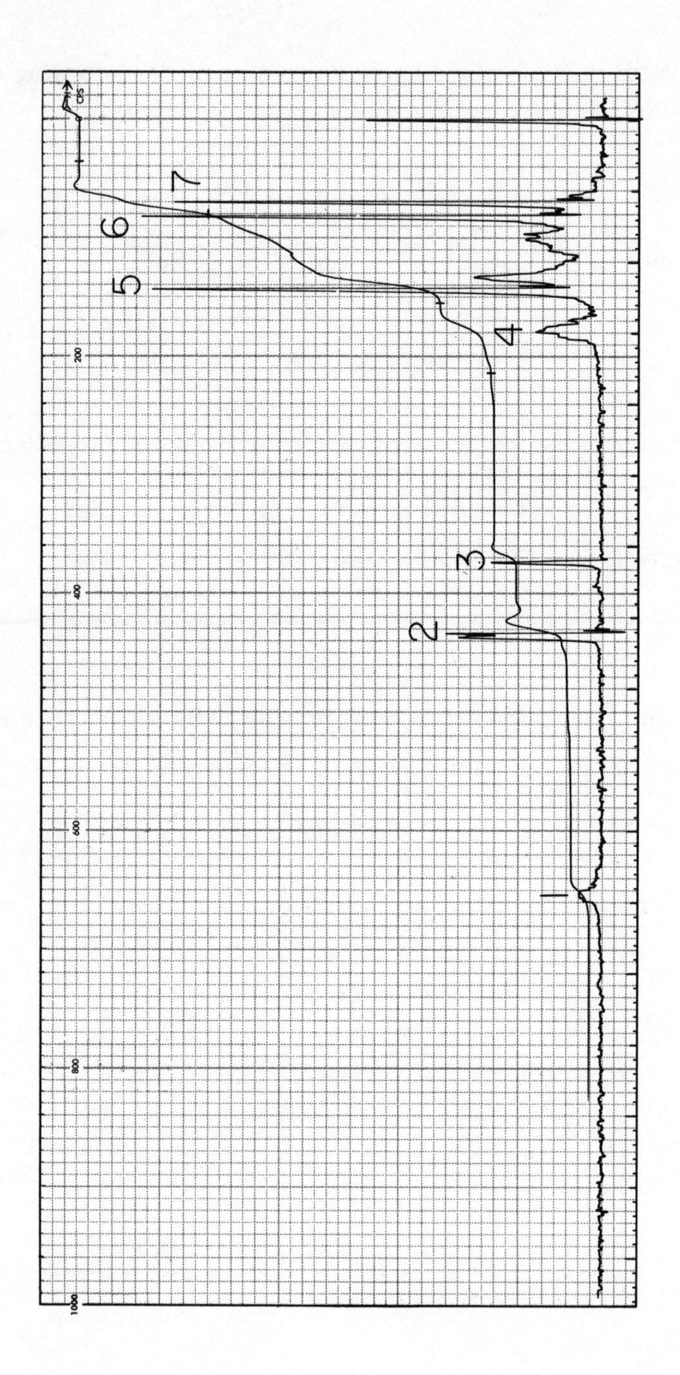
CPS
200
400
600
800
1000
1
2
3
4
5
6
7

This is the spectrum over the region 1000 to 0 cps of the following podocarpic acid derivative:

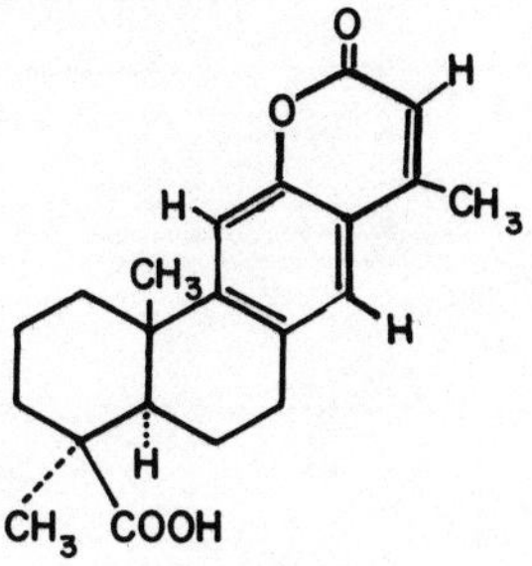

$C_{21}H_{24}O_4$

The spectrum after $D_2O$ exchange is shown on p. 58.

| Band | ① | ② | ③ | ④ | ⑥+⑦ | Total for spectrum |
|---|---|---|---|---|---|---|
| Integration | 7.2 | 21 | 9 | 21 | 54 | 216 |

---

*Questions*

1. Assign all of the numbered bands to the various groups of protons in the molecule.
2. What impurity should be expected to appear in band ②?

Answers: p. 218.

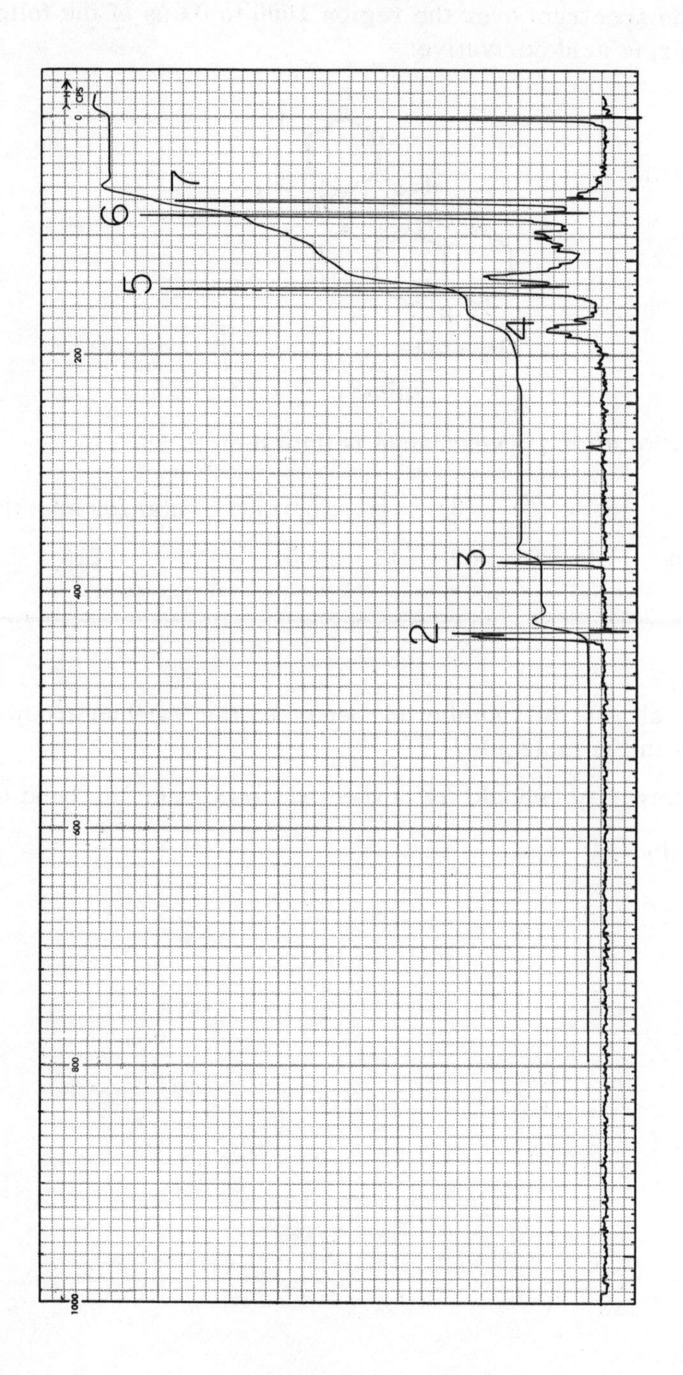

0 CPS
200
400
600
800
1000
2
3
4
5
6
7

This is the spectrum, after $D_2O$ exchange, of the compound used on p. 56. The region 1000 to 0 cps is shown.

---

*Questions*

1. Which peak is due to HOD?
2. Could the slight elevation in the integration curve near 220 cps be due to contamination with methanol?

Answers: p. 218.

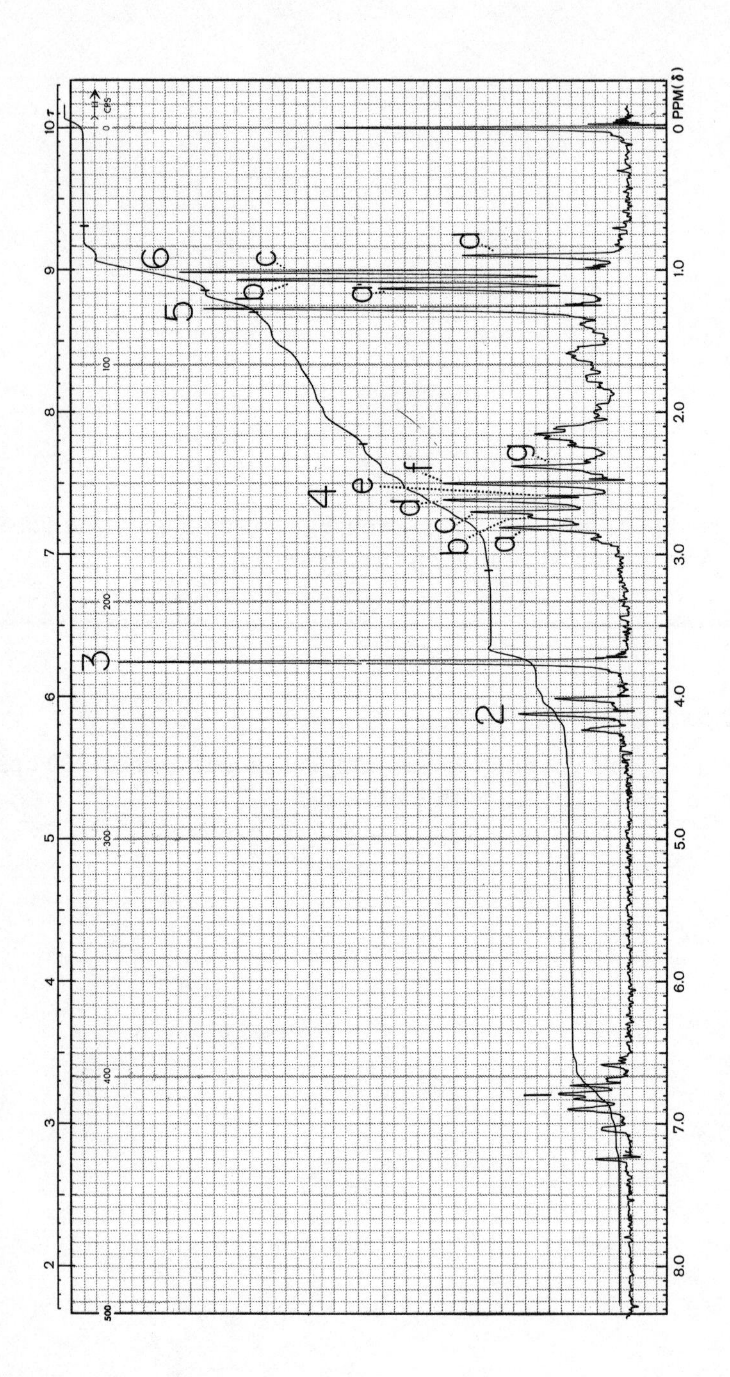
0 PPM(δ)
1.0
2.0
3.0
4.0
5.0
6.0
7.0
8.0
10 τ
9
8
7
6
5
4
3
2
CPS
100
200
300
400
500
1
2
3
4
5
6
a
b
c
d
e
f
g

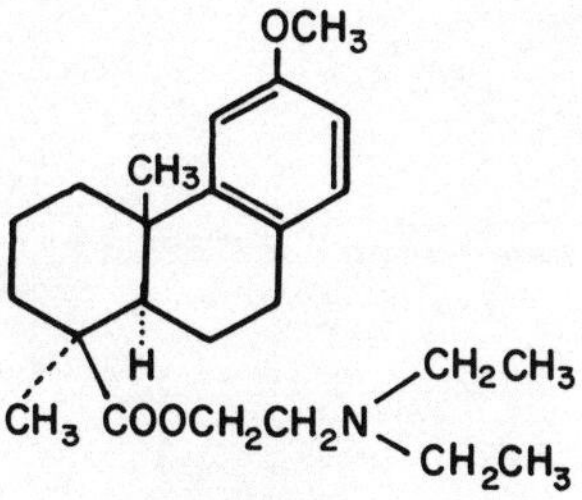

$C_{24}H_{37}NO_3$

| Band | ① | ② | ③ | ④ | ⑤ | ⑥ | Total for spectrum |
|---|---|---|---|---|---|---|---|
| Integration | 19.5 | 13 | 19 | 53 | 20 | 52 | 226 |

---

*Questions*

1. Assign all of the numbered bands and lettered peaks.
2. Describe the aromatic protons in notational terms.

Answers: p. 219.

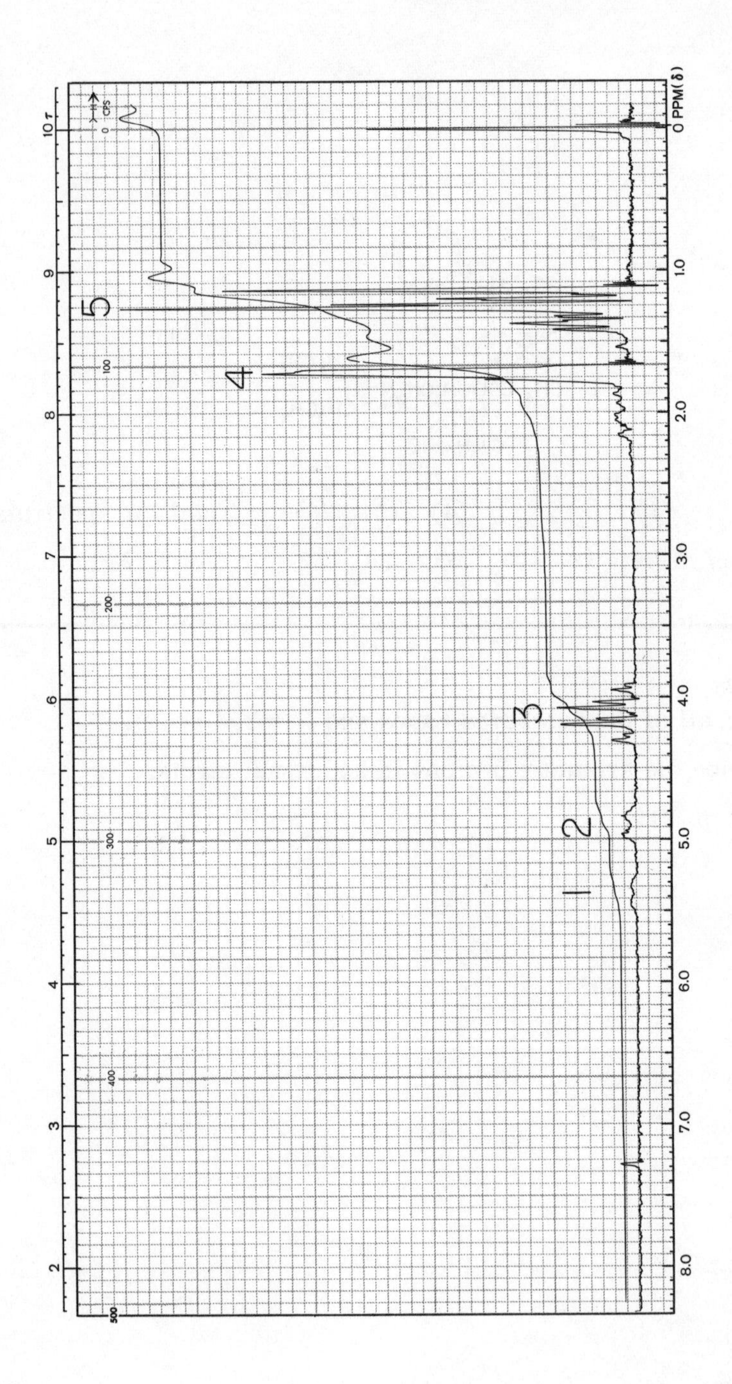

10 τ
9
8
7
6
5
4
3
2
0 PPM (δ)
1.0
2.0
3.0
4.0
5.0
6.0
7.0
8.0
0 CPS
100
200
300
400
500
1
2
3
4
5

This is a spectrum of a mixture of the following two isomers:[1]

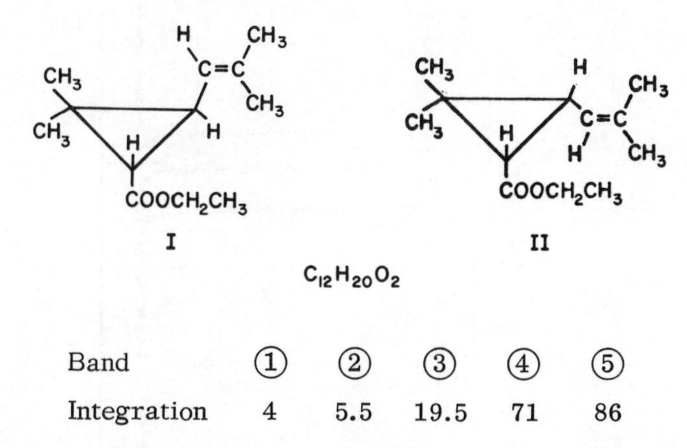


| Band | ① | ② | ③ | ④ | ⑤ |
|---|---|---|---|---|---|
| Integration | 4 | 5.5 | 19.5 | 71 | 86 |

---

*Questions*

1. Assign as many of the bands as possible.
2. What simple procedure might help in the assignment of the peaks in band ⑤?
3. What is the molar ratio of the two components in this mixture?

Answers: p. 219.

[1] A similar mixture of these two isomers is discussed by H. M. Hutton and T. Schaefer, Can. J. Chem. 40:875 (1962).

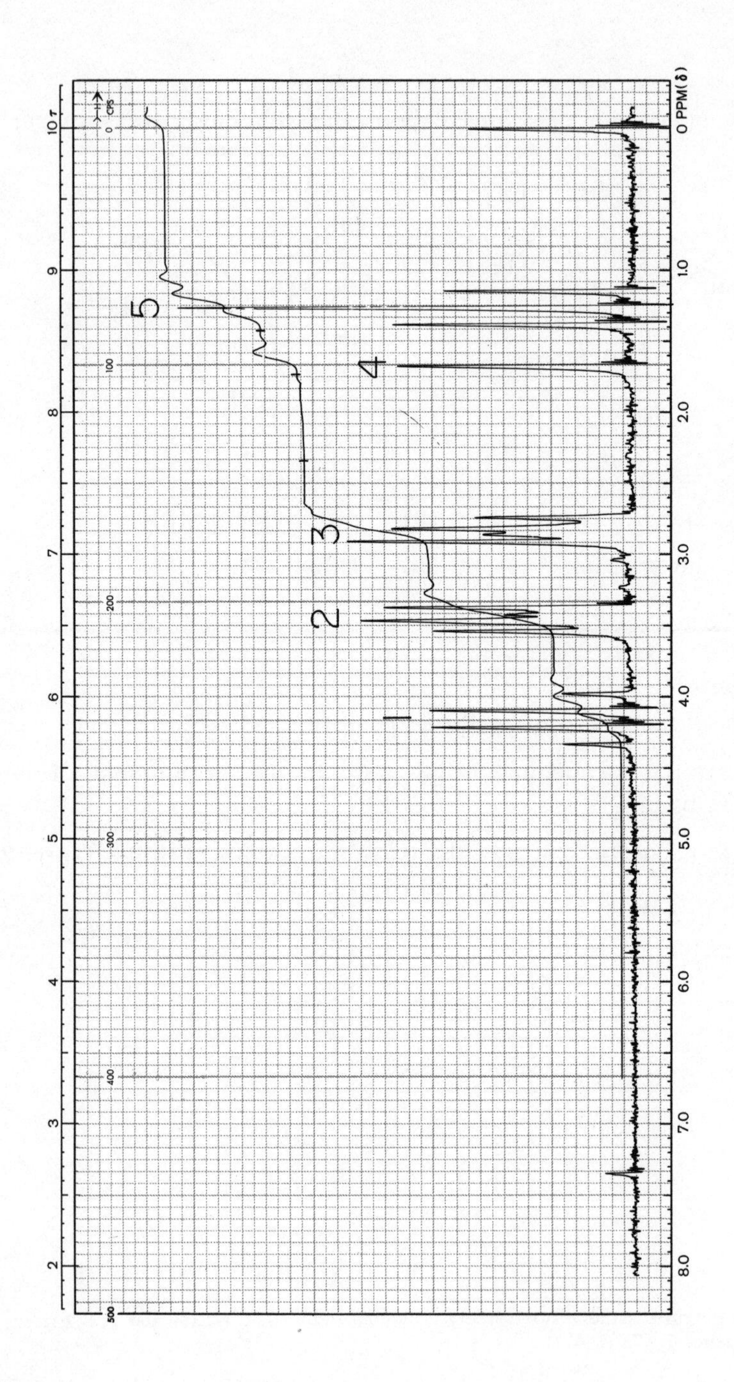
10 τ
9
8
7
6
5
4
3
2
0 PPM(δ)
1.0
2.0
3.0
4.0
5.0
6.0
7.0
8.0
0 CPS
100
200
300
400
500
1
2
3
4
5

$$HN\langle\rangle N{-}\overset{O}{\overset{\|}{C}}OCH_2CH_3$$

$C_7H_{14}N_2O_2$

The spectrum after $D_2O$ exchange is shown on p. 66.

| Band | ① | ② | ③ | ④ | ⑤ | Sum |
|---|---|---|---|---|---|---|
| Integration | 28 | 52.5 | 51.5 | 15.5 | 39 | 186.5 |

---

*Questions*

1. Assign all of the absorption bands to specific protons or groups of protons.
2. Classify the multiplets as first- or higher-order patterns.
3. How would these spin systems be designated?

Answers: p. 220.

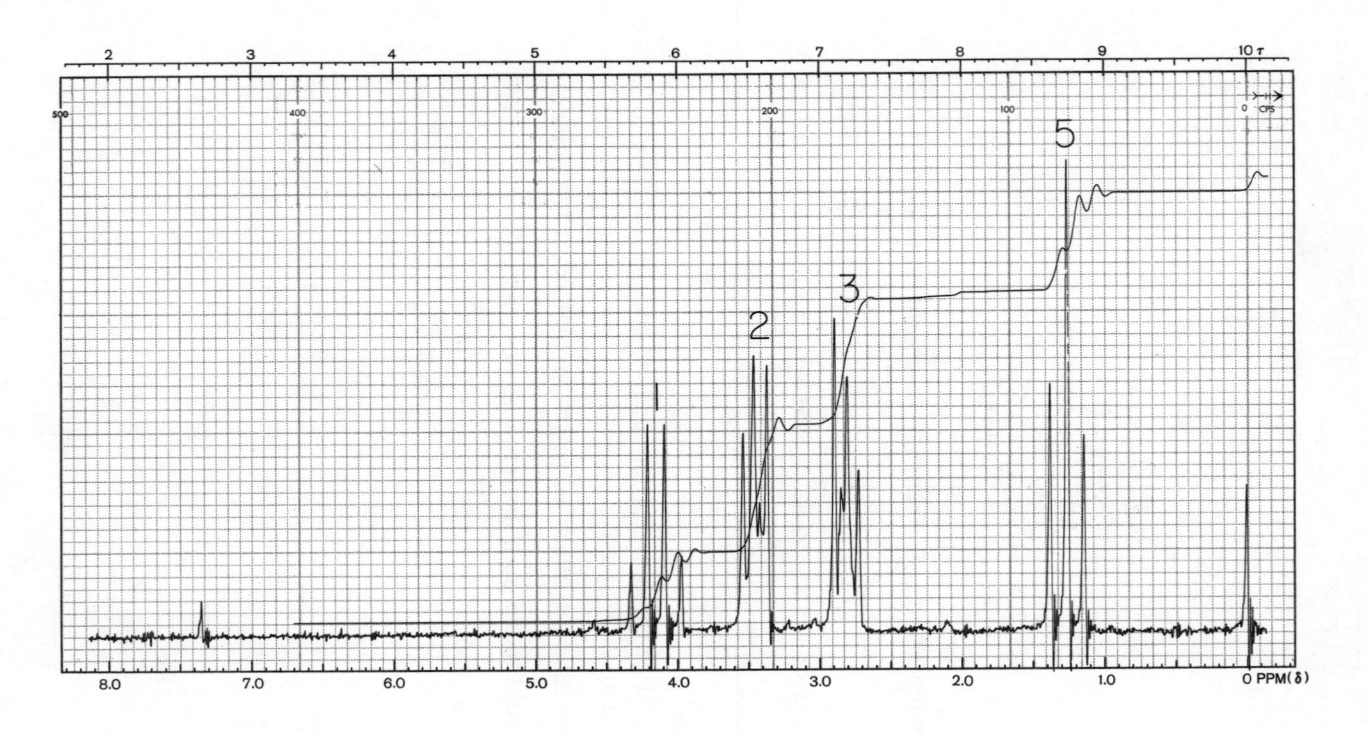

2
3
4
5
6
7
8
9
10 τ
500
400
300
200
100
0 CPS
5
3
2
1
8.0
7.0
6.0
5.0
4.0
3.0
2.0
1.0
0 PPM(δ)

$$DN\;\bigcirc\;N-\overset{O}{\overset{\|}{C}}OCH_2CH_3$$

$C_7H_{13}DN_2O_2$

This is the $D_2O$-exchange spectrum corresponding to the spectrum on p. 64.

---

*Question*

1. What information can be derived from the distortions of the multiplets from the predicted ratios (1:1, 1:2:1, etc.)?

Answer: p. 221.

# SECTION 4

First-order couplings to $H^2$, $C^{13}$, $F^{19}$, and $P^{31}$; spinning side bands.

*Suggested Reading*

Bhacca and Williams: pp. 123-132.

Bible: pp. 59-65.

Emsley, Feeney, and Sutcliffe: pp. 183-197, 879-1029, and 1061-1068.

Pople, Schneider, and Bernstein: pp. 162-164, 195-198, and 400-402.

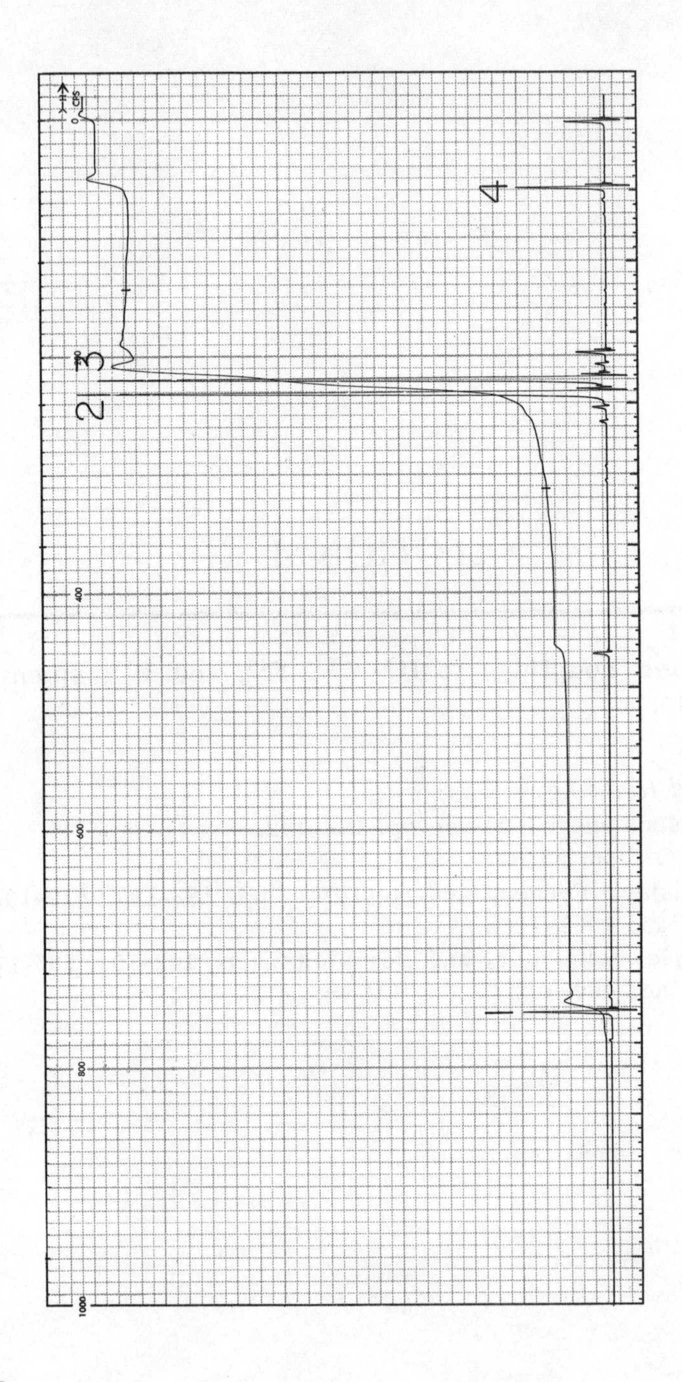

This spectrum of $HP(=O)(OCH_3)_2$ was determined over the region 1000 to 0 cps using the pure liquid.

| Band | ① | ②+③ | ④ |
|---|---|---|---|
| Integration | 14.5 | 175 | 13 |

---

*Questions*

1. Assign the numbered peaks.
2. Explain the splittings.
3. What is the chemical shift of the hydrogen attached to the phosphorus?

Answers: p. 221.

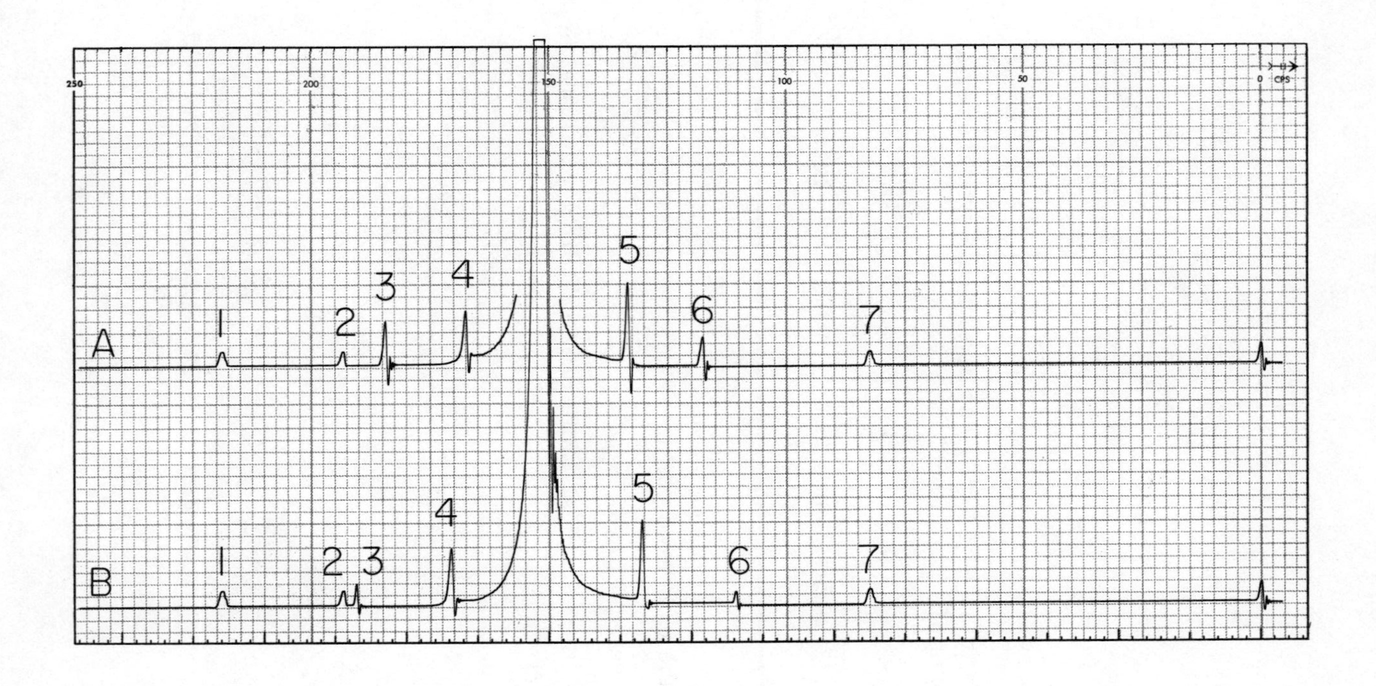
250
200
150
100
50
0 CPS
A
1
2
3
4
5
6
7
B
1
2 3
4
5
6
7

$$CH_3\overset{\overset{O}{\|}}{S}CH_3$$

$$C_2H_6OS$$

A small amount of TMS was added to a sample of dimethylsulfoxide. The spectrum of this mixture was then determined at two different sample spinning rates, using a relatively high gain, over the region from 250 to 0 cps. The signal near 152 cps, which is due to the six protons in the two methyl groups, was allowed to go off scale.

---

*Questions*

1. Identify the spinning side bands.
2. Was the sample spinning faster during the determination of scan A or scan B ?
3. Explain the origin of the other bands.

Answers: p. 222.

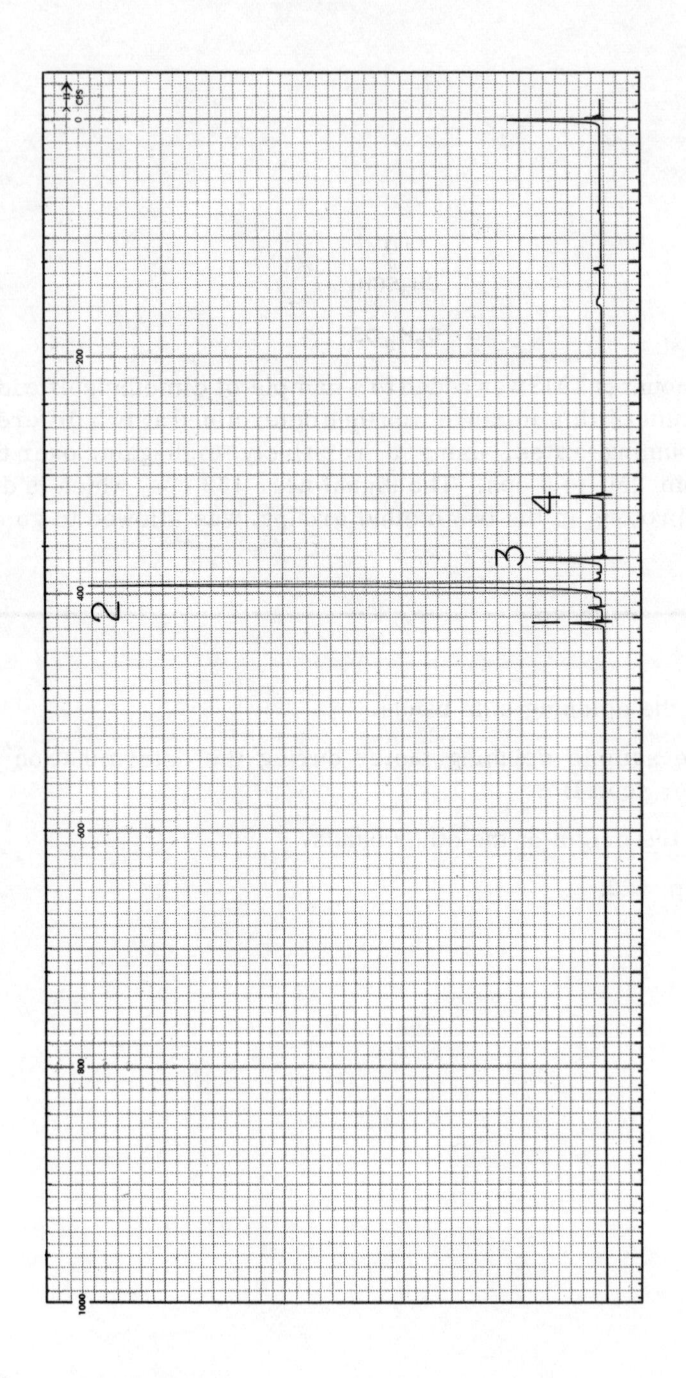
0 CPS
200
400
600
800
1000
1
2
3
4

$CHF_2COOH + D_2O$

A spectrum of difluoroacetic acid in deuterodimethylsulfoxide had peaks ① and ② superimposed. A few drops of $D_2O$ were added. The spectrum that was obtained on this diluted sample over the region 1000 to 0 cps is shown on the left.

---

*Questions*

1. Assign the peaks to specific protons.
2. Cite several factors which are probably involved in the shift of the carboxylic acid-proton signal on dilution with $D_2O$.

Answers: p. 222.

# SECTION 5

Simple unexpected nonequivalence of protons.

*Suggested Reading*

Bible: pp. 65-75.
Emsley, Feeney, and Sutcliffe: pp. 551-581 and 1037-1041.
Jackman: pp. 99-103.
Pople, Schneider, and Bernstein: pp. 365-386.
Roberts: pp. 57-60 and 69-86.

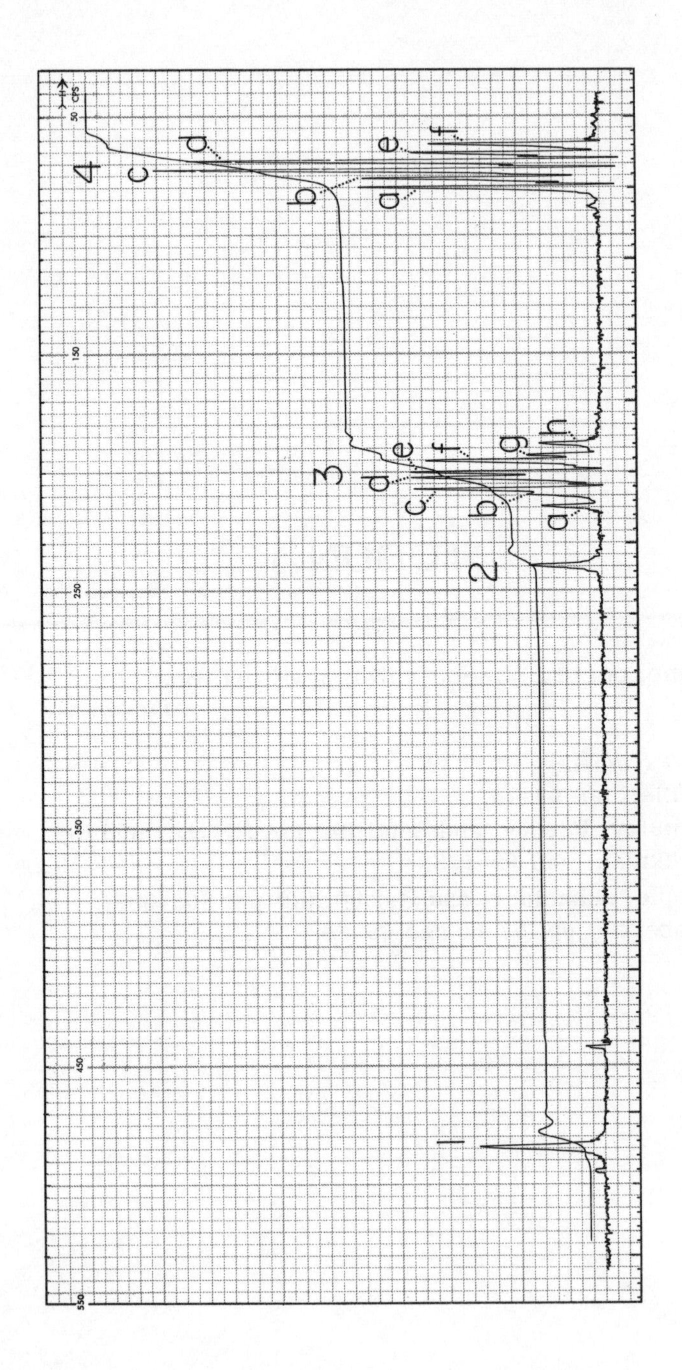

CPS
50
150
250
350
450
550
1
2
3
4

$$HC(=O)N(CH_2CH_3)_2$$

$C_5H_{11}NO$

The spectrum of the above compound is offset 50 cps.

| Band | ① | ② | ③ | ④ |
|---|---|---|---|---|
| Integration | 17 | 10 | 70 | 107 |

---

*Questions*

1. Assign all of the bands.
2. Explain the multiplicities in bands ③ and ④.

Answers: p. 223.

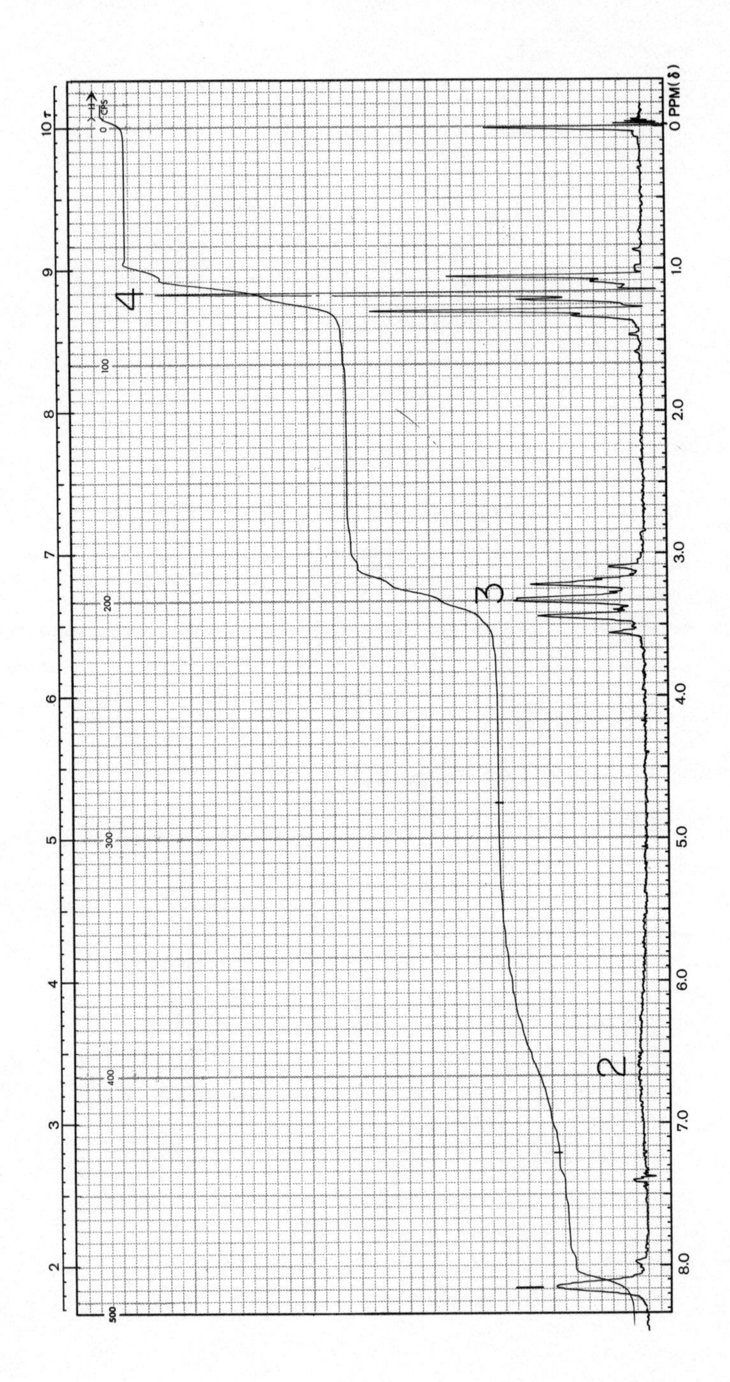

10 τ
9
8
7
6
5
4
3
2
0 CPS
100
200
300
400
500
0 PPM (δ)
1.0
2.0
3.0
4.0
5.0
6.0
7.0
8.0
1
2
3
4

$$\overset{O}{\overset{\|}{HC}}NHCH_2CH_3$$

$C_3H_7NO$

| Band | ① | ② | ③ | ④ |
|---|---|---|---|---|
| Integration | 26 | 24 | 61 | 91 |

The spectrum determined after a 10-min $D_2O$ exchange is given on p. 84, while the spectrum determined after $D_2O$ exchange for several hours is given on p. 82.

---

*Questions*

1. Assign all of the bands.
2. Explain the appearance of the band due to the —NH— proton.
3. Explain the splitting of band ③.
4. Suggest possible explanations for the spikes on the low-field (left) side of the triplet in band ④.
5. What other factor could complicate this spectrum even more?

Answers: p. 223.

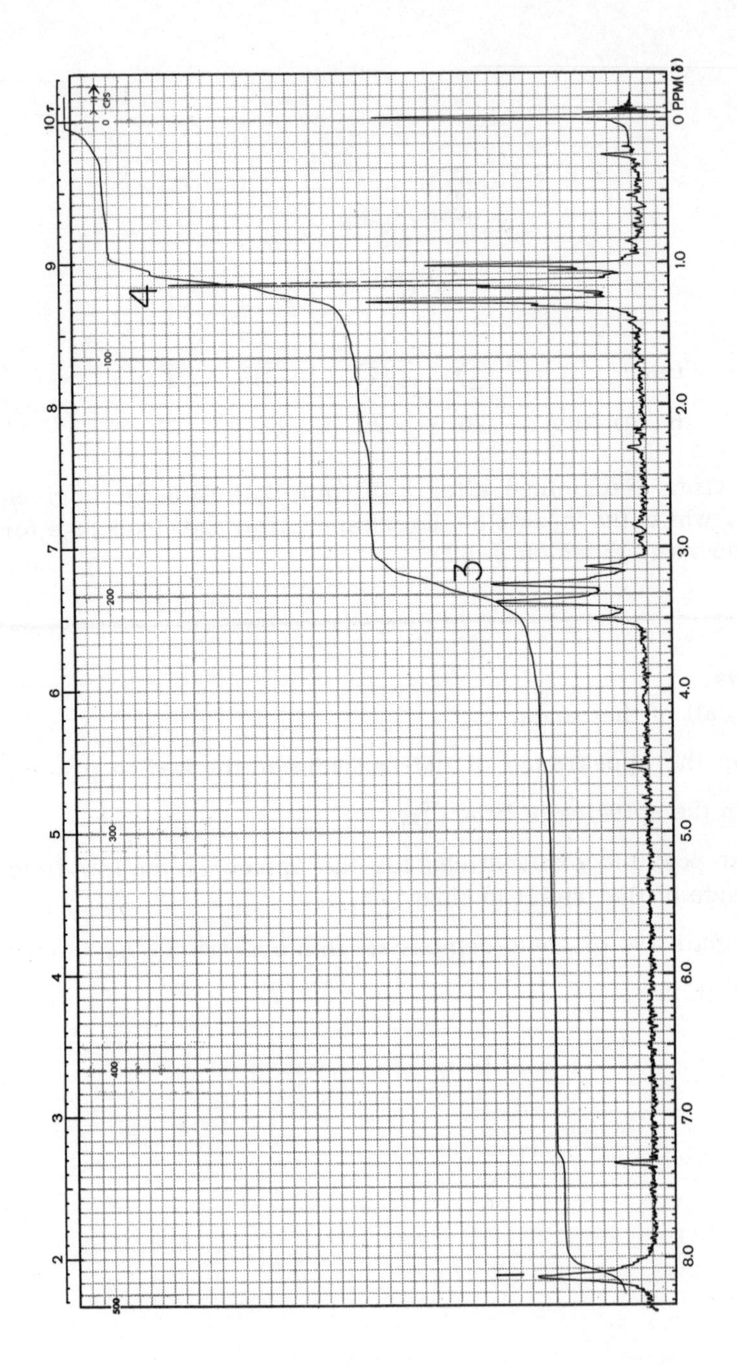

PPM(δ)
CPS
10 τ

$$\overset{O}{\overset{\|}{HC}}NDCH_2CH_3$$

$C_3H_6DNO$

After the spectrum on p. 80 was determined, the $CDCl_3$ solution was shaken with a few drops of $D_2O$ and then allowed to stand for several hours. The resulting solution gave the spectrum on the left. The amplifier gain had to be increased for this spectrum because some of the ethyl formamide was washed out of the $CDCl_3$ solution. The amplifier gain then had to be reduced to keep the TMS signal on scale. This accounts for the lower noise level in the 0 to 10-cps region.

---

*Questions*

1. Explain the difference in the appearance of the band due to the methylene protons before and after $D_2O$ exchange.

2. Suggest a reason for the broadness of band ①.

Answers: p. 224.

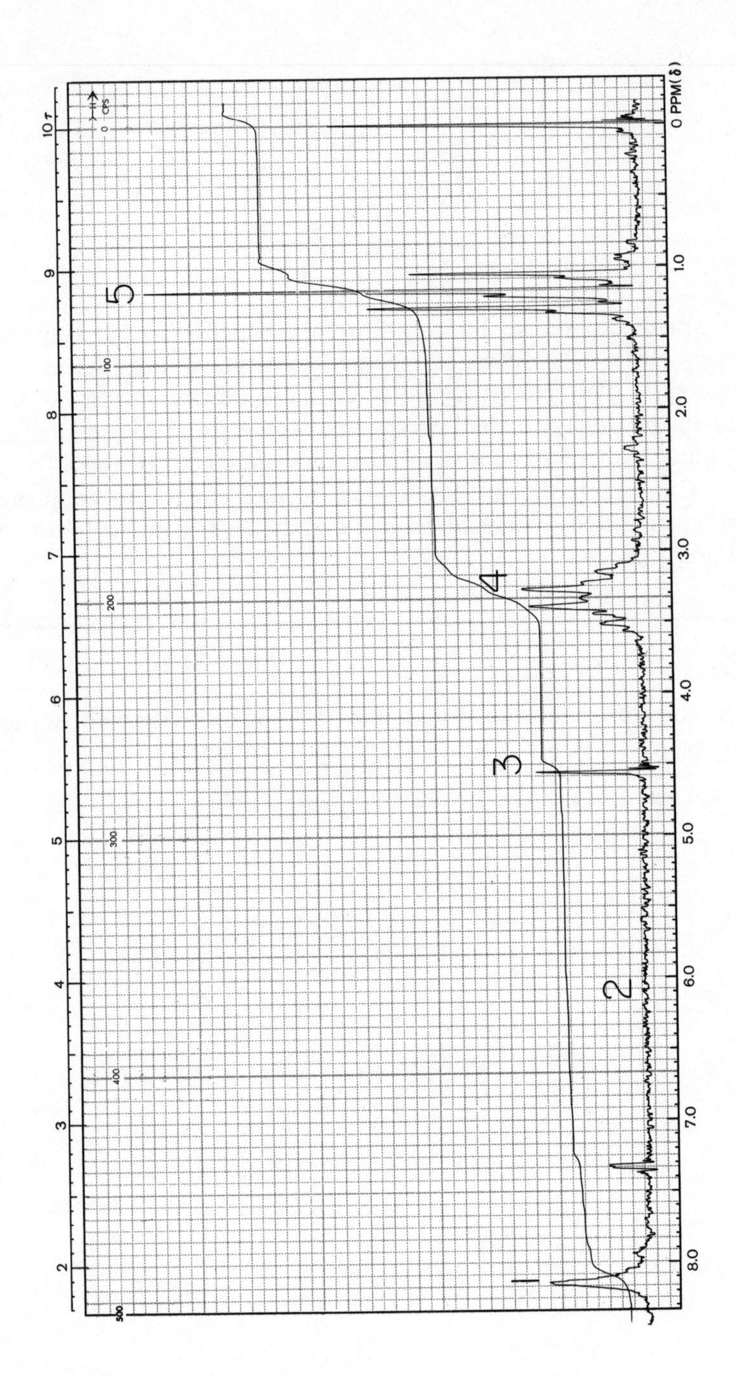

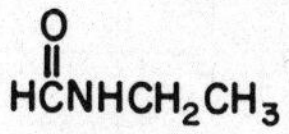

$$\overset{\overset{\large O}{\|}}{H C} N H C H_2 C H_3$$

$C_3H_7NO$

This spectrum of ethyl formamide was determined after a 10-min $D_2O$ exchange.

---

*Questions*

1. Identify band ③.
2. Explain the appearance of band ④.

Answers: p. 225.

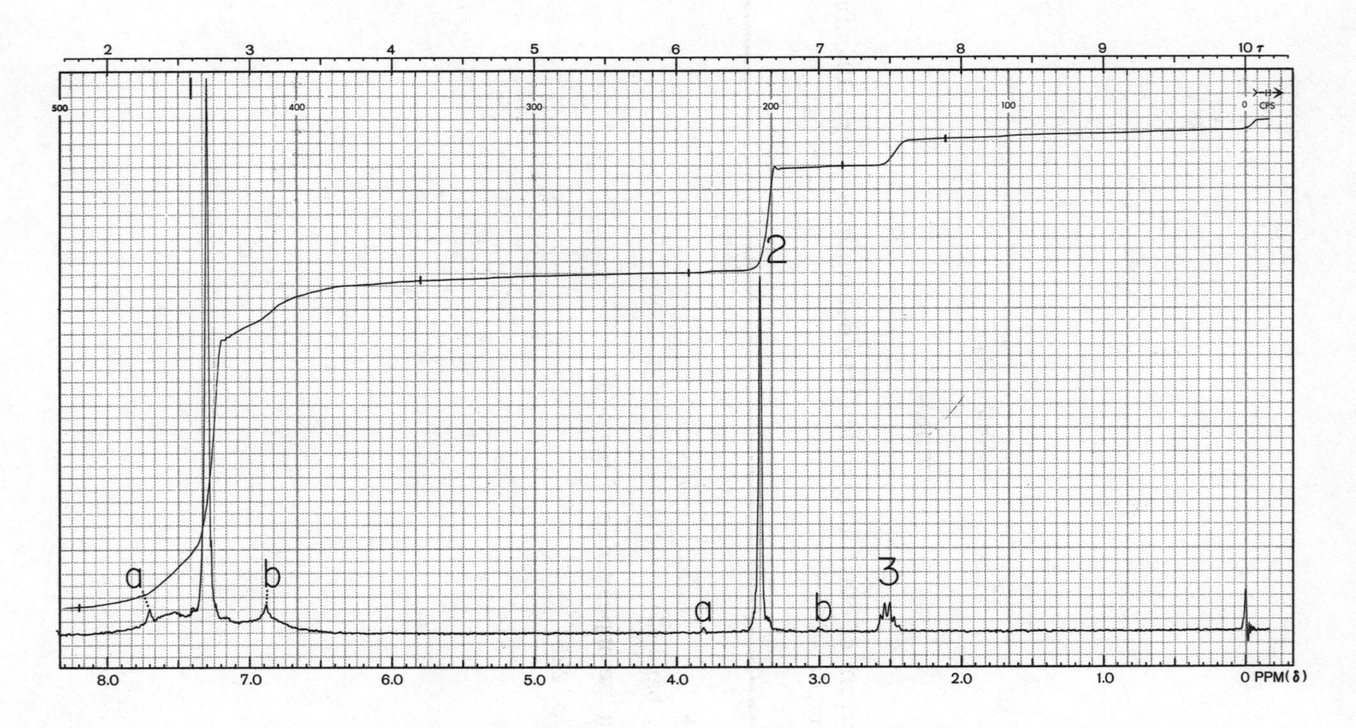
2
3
4
5
6
7
8
9
10 τ
500
400
300
200
100
0
CPS
1
2
3
a
b
a
b
8.0
7.0
6.0
5.0
4.0
3.0
2.0
1.0
0 PPM(δ)

$$\text{C}_6\text{H}_5\text{-CH}_2\overset{\overset{\displaystyle O}{\|}}{\text{C}}\text{NH}_2$$

$C_8H_9NO$

This spectrum was determined in $CD_3\overset{\overset{O}{\|}}{S}CD_3$ using TMS as an internal reference.

| Band | ① | ② | ③ |
|---|---|---|---|
| Integration | 137 | 45 | 12 |

---

*Questions*

1. Identify each numbered band and lettered peak.
2. The phenyl protons are expected to be coupled with each other. Why is only a single peak observed?

Answers: p. 225.

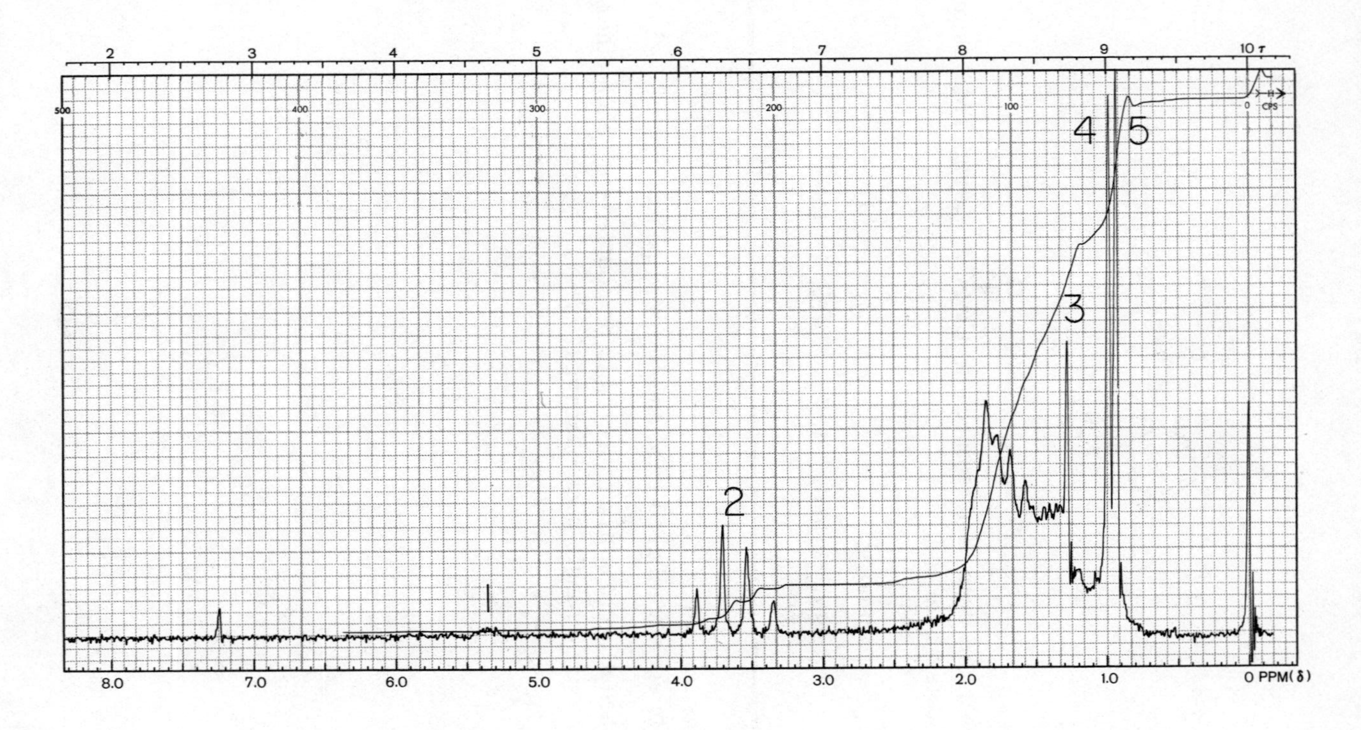

2
3
4
5
6
7
8
9
10 τ
500
400
300
200
100
0
CPS
1
2
3
4
5
8.0
7.0
6.0
5.0
4.0
3.0
2.0
1.0
0 PPM(δ)

This material could have the following structures:

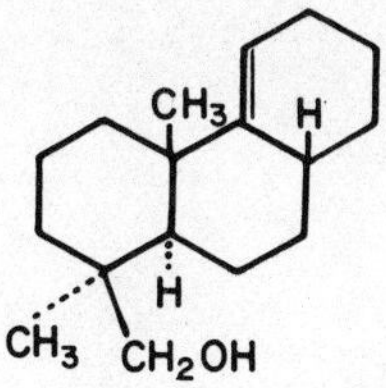

I 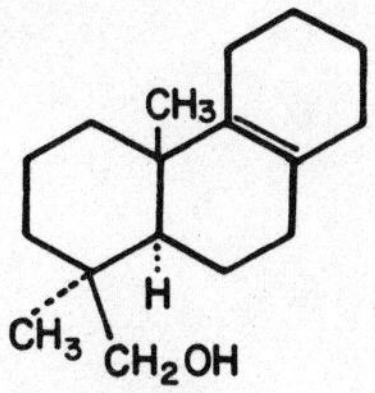     II 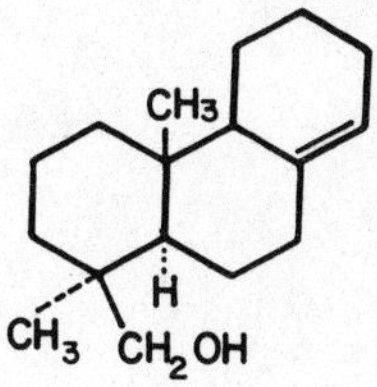     III

$C_{17}H_{28}O$

Neither exchange with $D_2O$ nor sweeping from higher to lower field alters the appearance of band ②. Exchange with $D_2O$ removes peak ③.

---

*Questions*

1. Assign the numbered bands.
2. Which is the correct structure?
3. Explain the splitting in band ②.
4. Is the coupling constant involved in band ② consistent with the assignment of this band?
5. Suggest an explanation of the slight broadening of the high-field doublet in band ②.

Answers: p. 225.

# SECTION 6

Higher-order spin systems; AB, $AB_2$, $A_2B_2$, $A'_2B'_2$, ABX, and ABC; calculations for AB, $AB_2$, and ABX; "virtual" coupling.

*Suggested Reading*

Bhacca and Williams: pp. 42-47, 77-78, 135-150, and 176-181.
Bible: pp. 77-104.
Emsley, Feeney, and Sutcliffe: pp. 311-428 and 794-816.
Jackman: pp. 24-26 and 88-98.
Pople, Schneider, and Bernstein: pp. 119-151.
Roberts: pp. 48-53 and 55-57.

J. D. Roberts in the text, An Introduction to the Analysis of Spin–Spin Splitting in High-Resolution NMR Spectra, W. A. Benjamin, Inc. (New York), 1961, presents a simplified theory of both first- and higher-order spin–spin interactions.

Wiberg, K. B., and B. J. Nist, The Interpretation of NMR Spectra, W. A. Benjamin, Inc. (New York), 1962. In particular, the introductions to the AB, $AB_2$, ABX, and ABC cases (pp. 3, 11-12, and 21-29) bear directly on some of the questions that appear in this section.

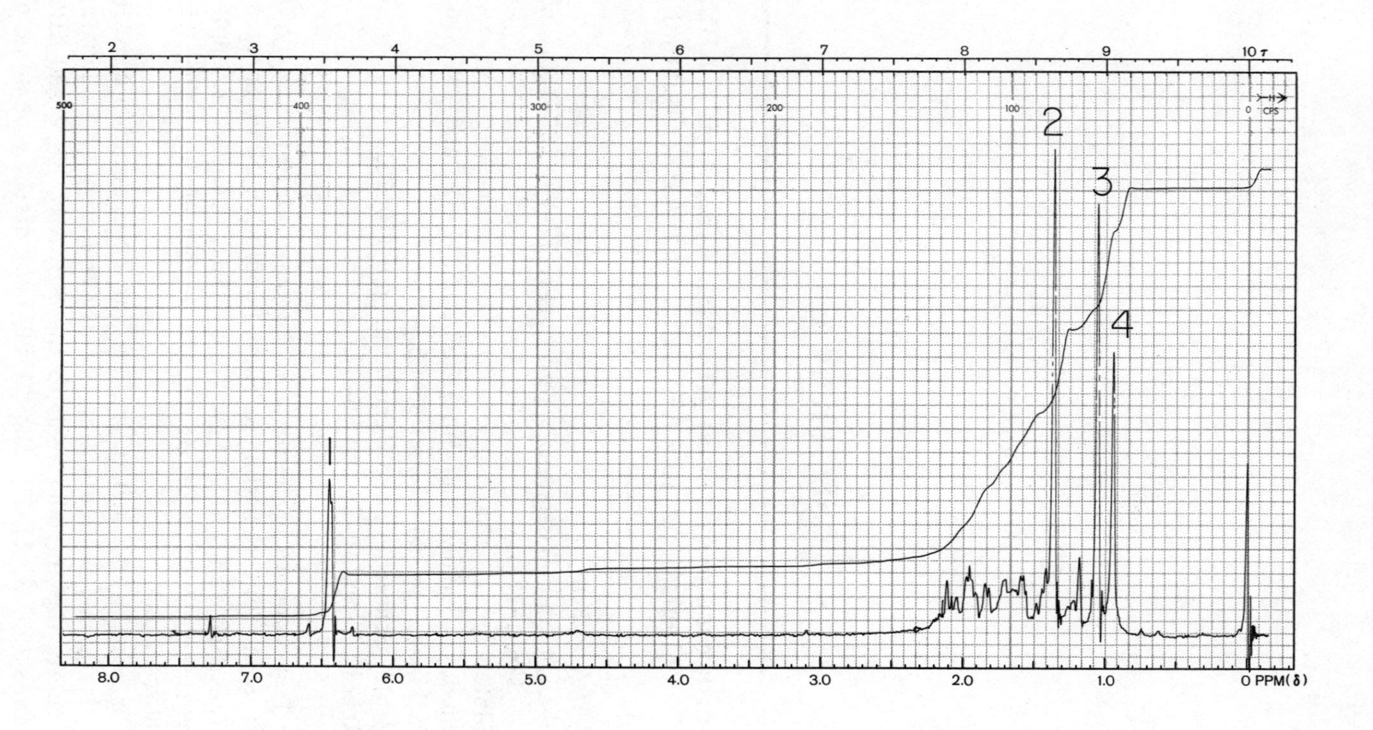

2 3 4 5 6 7 8 9 10 τ
500 400 300 200 100 0 CPS
1
2
3
4
8.0 7.0 6.0 5.0 4.0 3.0 2.0 1.0 0 PPM(δ)

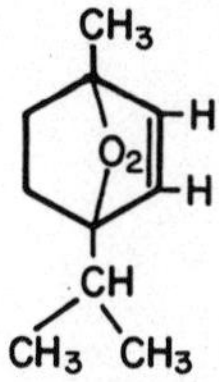

$C_{10}H_{16}O_2$

| Band | ① | ③ + ④ | Total for spectrum |
| --- | --- | --- | --- |
| Integration | 17 | 47 | 182 |

*Questions*

1. Assign the numbered peaks to specific protons.
2. Explain the appearance of band ①. An expansion of this band is shown on p. 94.

Answers: p. 226.

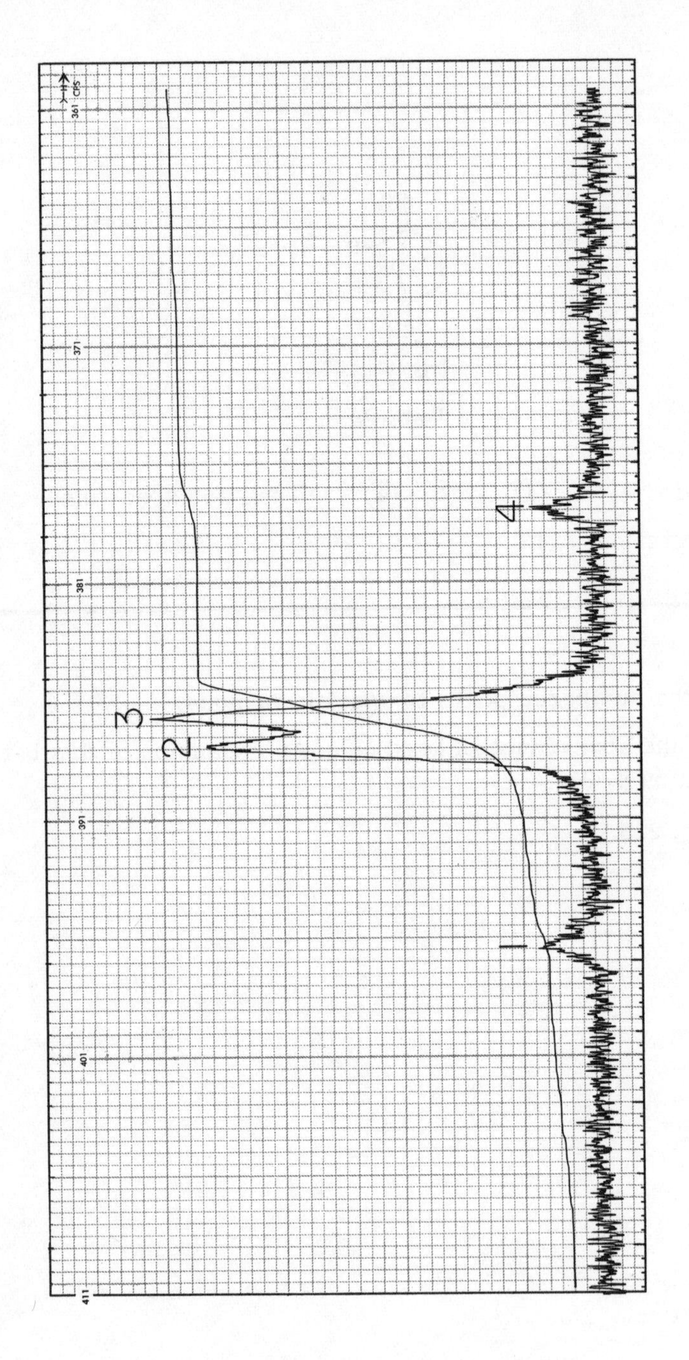

361 CPS
371
381
391
401
411
1
2
3
4

This is an expansion of the olefinic proton signals of the following compound:

The region 411 to 361 cps (6.71 to 6.01 ppm) is shown. The complete proton spectrum is given on p. 92.

| Peak | ① | ② | ③ | ④ |
|---|---|---|---|---|
| Position (cps) | 396.6 | 387.9 | 386.7 | 378.0 |
| Integrated intensity | 8 | Total: | 139 | 7 |

---

*Questions*

1. For this AB pattern, determine $J_{AB}$, $\Delta\nu_{AB}$, and the chemical shifts of the two olefinic protons.
2. Compare the calculated difference in chemical shifts between A and B with the first-order approximation.
3. Using average values for the intensities, compare the observed and calculated ratios of the intensities of the smaller to the larger peaks.

Answers: p. 227.

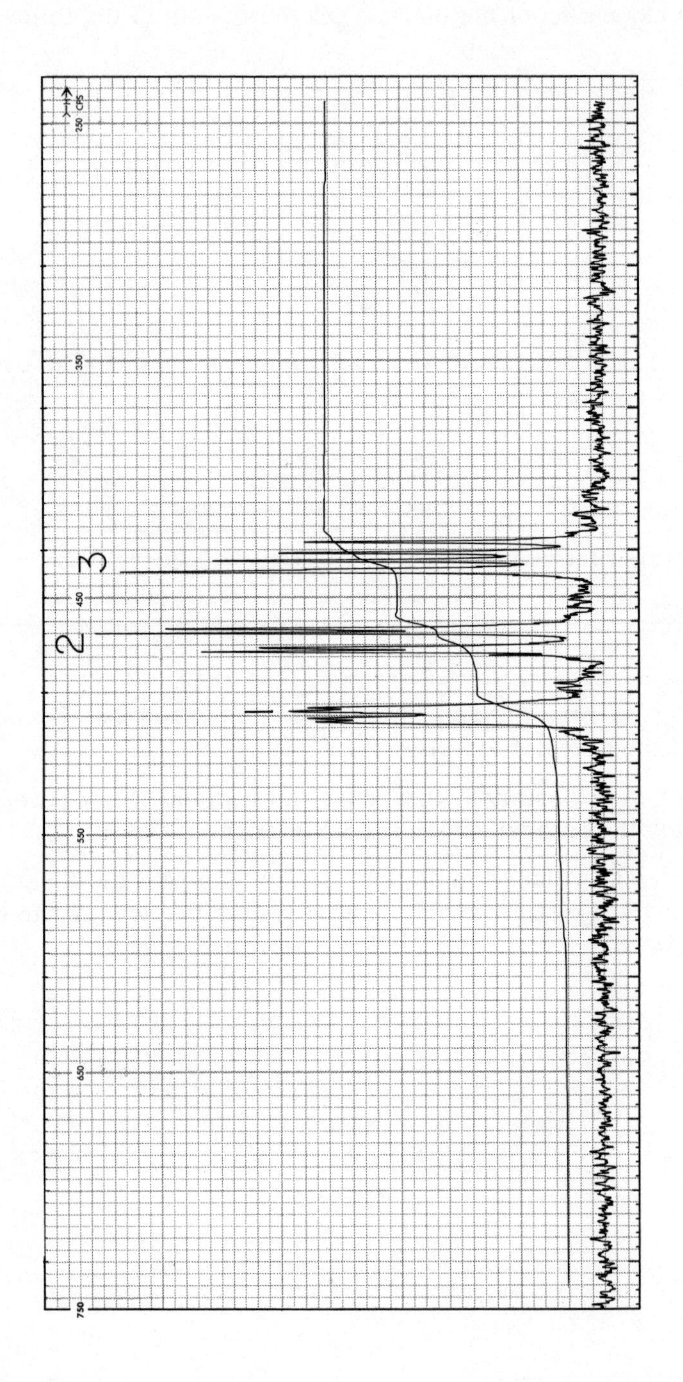
250 CPS
350
450
550
650
750
1
2
3

This spectrum was determined over the region 750 to 250 cps (offset of 250 cps with a sweep width of 500 cps). The compound used was one of the following six isomers:

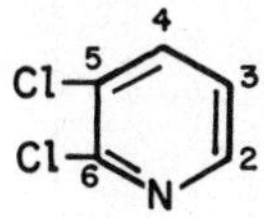

I 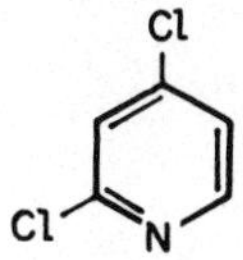 II 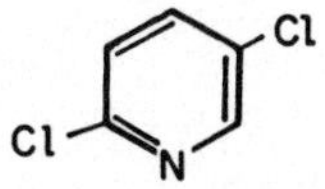 III

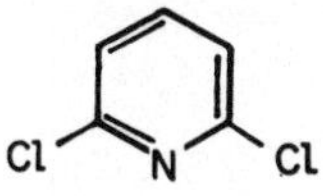

IV 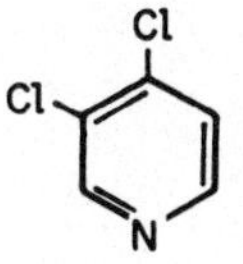 V 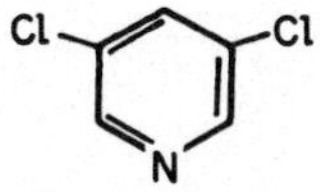 VI

$C_5H_3Cl_2N$

---

*Questions*

1. To what type of spin system is this pattern due?
2. Which two of the six isomers would, of necessity, give some other pattern?
3. By assuming that this pattern is approximately an AMX type, estimate the three coupling constants.
4. Using the estimated coupling constants, choose the most likely isomer from the remaining four possibilities.
5. On the basis of the splittings, assign the bands to specific protons.

Answers: p. 227.

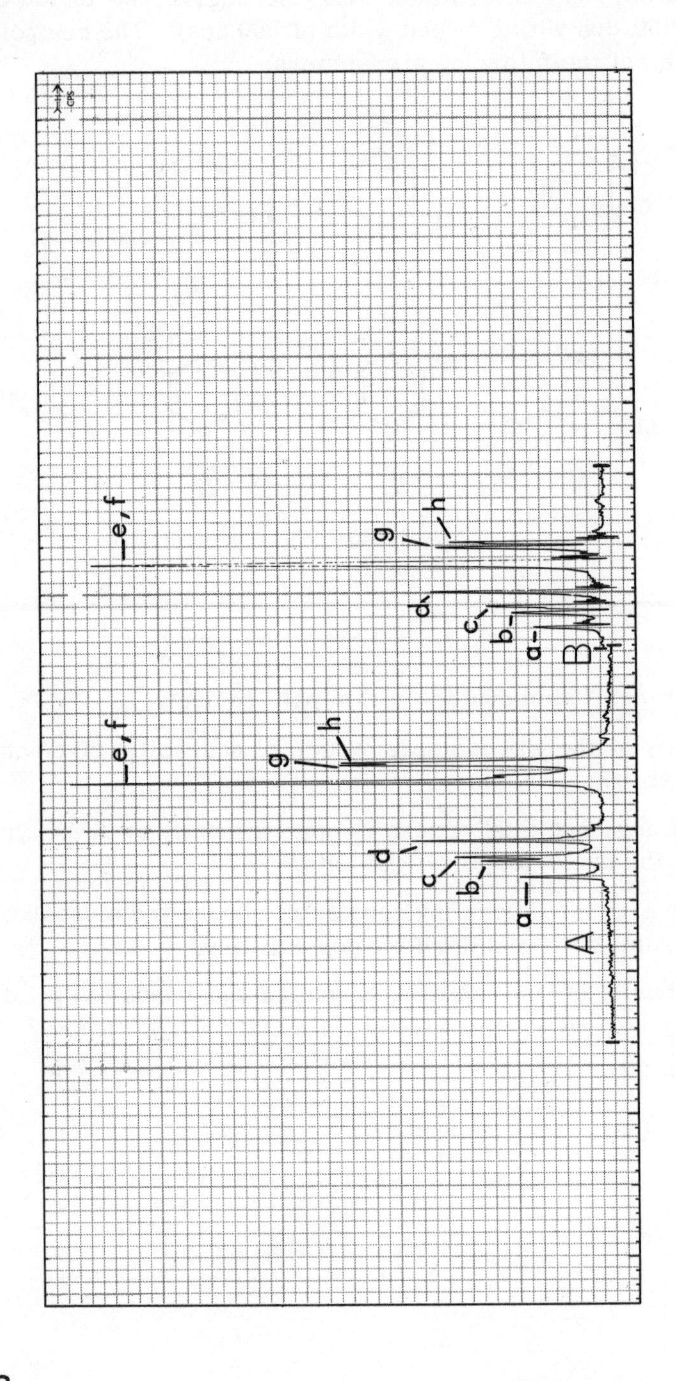

CPS
e, f
g
h
d
c
b
a
A
e, f
g
h
d
c
b
a
B

These two spectra were both run using the same compound. One of the scans was determined at 60 Mcps, while the other was determined at 100 Mcps. The scale in cps is the same for both scans (one scale division is equal to 5 cps). The compound used was another one of the following six possible dichloropyridines:

I II III

IV V VI

$C_5H_3Cl_2N$

---

*Questions*

1. Describe these patterns in notational terms.
2. Which of the six isomers was used?
3. Which spectrum was determined at the higher frequency?
4. Using the patterns given by Corio[1] or by Wiberg and Nist,[2] estimate the ratio $\Delta \nu_{AB}/J_{AB}$ in scan B.

Answers: p. 228.

[1] P.L. Corio, Chem. Rev. 60:363 (1960).
[2] K.B. Wiberg and B.J. Nist, The Interpretation of NMR Spectra, W.A. Benjamin, Inc. (New York), 1962, pp. 14-19.

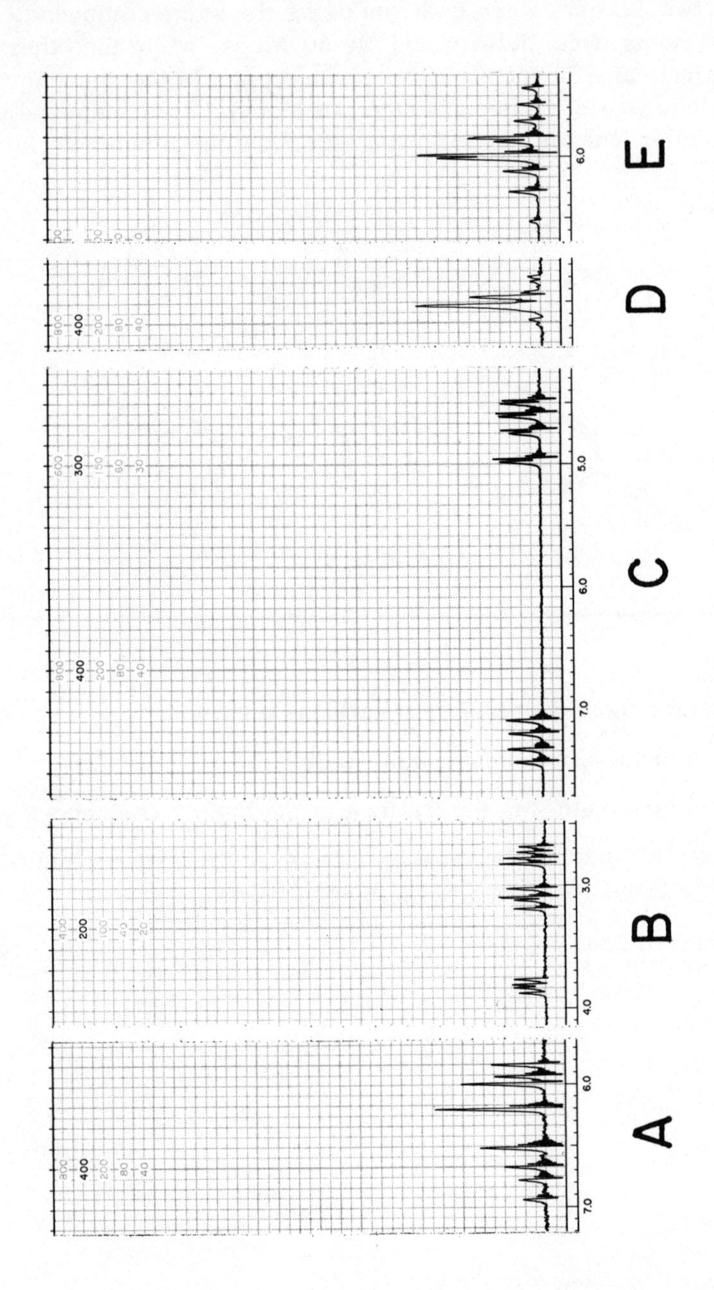
A
B
C
D
E

These three-spin patterns are reproduced from the Varian Catalogs[1] by permission of Varian Associates.

---

*Question*

1. Describe these patterns in notational terms.

Answer: p. 230.

[1] N.S. Bhacca, L.F. Johnson, and J.N. Shoolery, NMR Spectra Catalog, Vol. 1, Varian Associates (Palo Alto, California), 1962; and N.S. Bhacca, D.P. Hollis, L.F. Johnson, and E.A. Pier, NMR Spectra Catalog, Vol. 2, Varian Associates (Palo Alto, California), 1963.

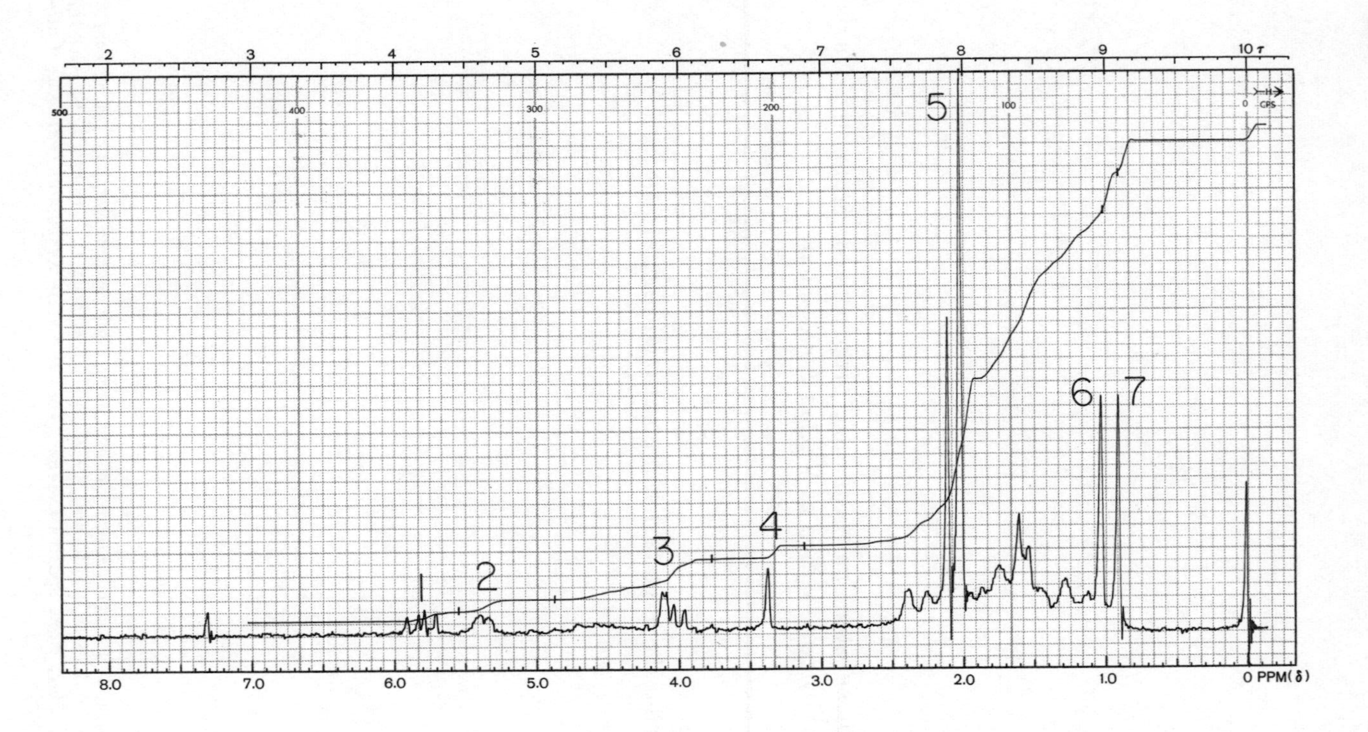
10 τ
9
8
7
6
5
4
3
2
0 PPM (δ)
1.0
2.0
3.0
4.0
5.0
6.0
7.0
8.0
0
100
200
300
400
500
CPS
1
2
3
4
5
6
7

$C_{27}H_{38}O_7$

An expansion of band ③ and one of bands ① and ② are shown on pp. 104 and 106, respectively.

| Band | ① | ② | ③ | ④ | ⑥ | ⑦ | Total for spectrum |
|---|---|---|---|---|---|---|---|
| Integration | 4 | 5.5 | 16.5 | 5.5 | 16 | 13.5 | 200.5 |

---

*Questions*

1. Assign the numbered bands to specific protons.
2. Explain the sharpness of band ④.
3. To what spin system do the sharp peaks in bands ① and ③ belong?

Answers: p. 231.

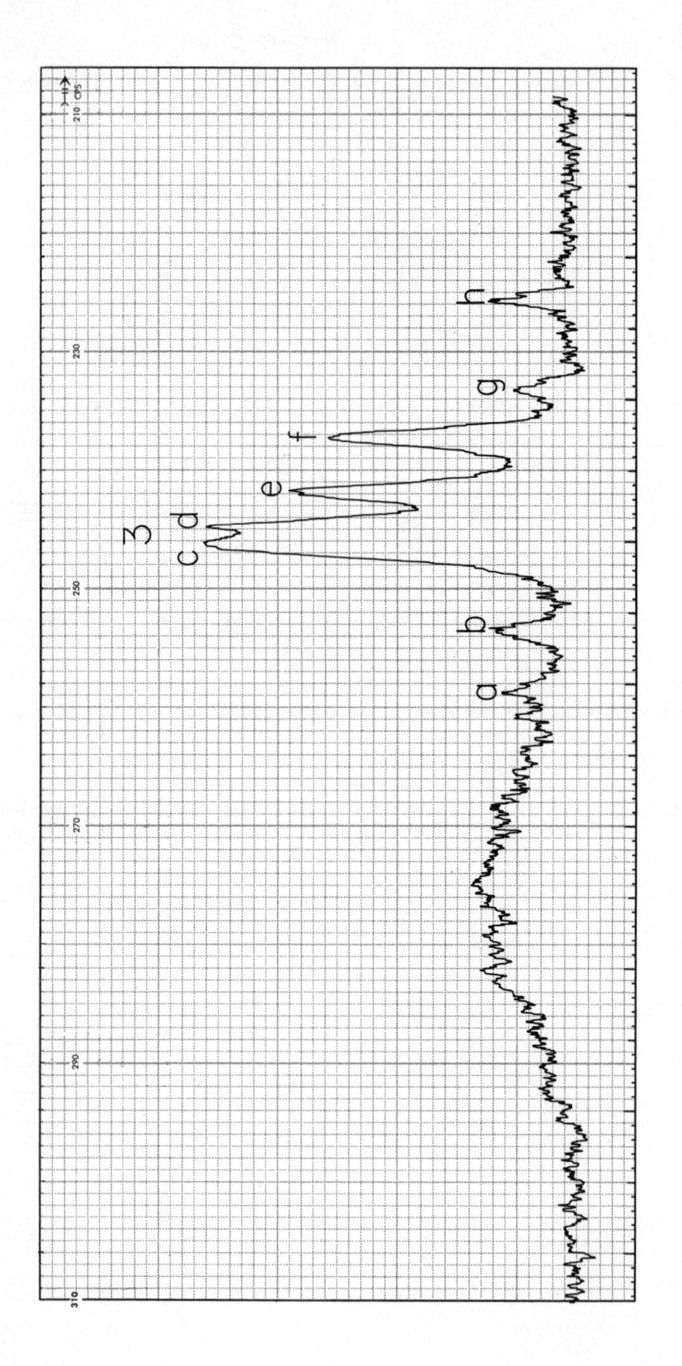
210 CPS
230
250
270
290
310
3
c
d
e
f
g
h
a
b

This is an expansion of band ③ in the spectrum on p. 102. The region 310 to 210 cps is shown. The peak positions can be taken as follows:

| Peak | Position |
|---|---|
| a | 258.2 |
| b | 253.8 |
| c | 246.6 |
| d | 245.1 |
| e | 242.1 |
| f | 237.6 |
| g | 233.6 |
| h | 226.0 |

The above information is needed for the answer to the question on p. 107.

---

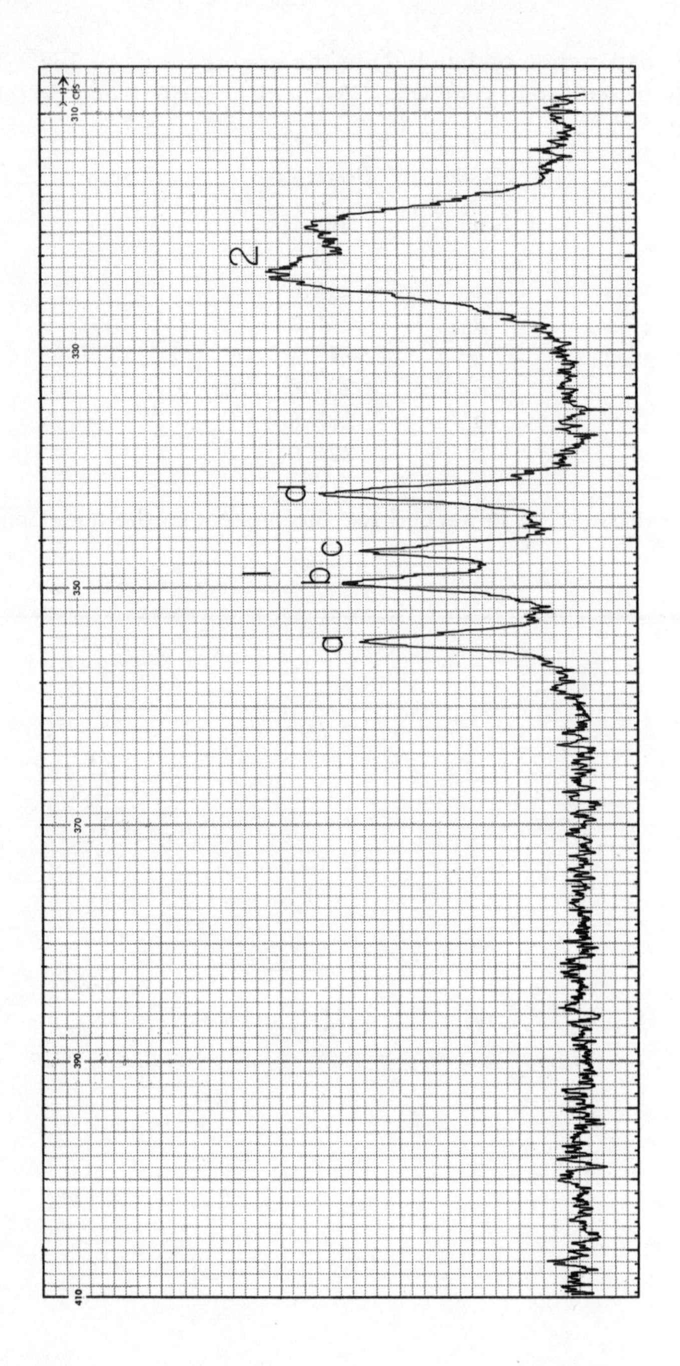

310 CPS
330
350
370
390
410
2
1
a
b c
d

This is the expansion of bands ① and ② in the spectrum on p. 102. The region 410 to 310 cps is shown. The peak positions in band ① can be taken as follows:

| Peak | Position |
|---|---|
| a | 354.5 |
| b | 349.6 |
| c | 346.8 |
| d | 342.0 |

---

*Question*

1. Using the numerical peak positions given above and on p. 105, analyze the ABX pattern seen in bands ① and ③.

Answer: p. 232.

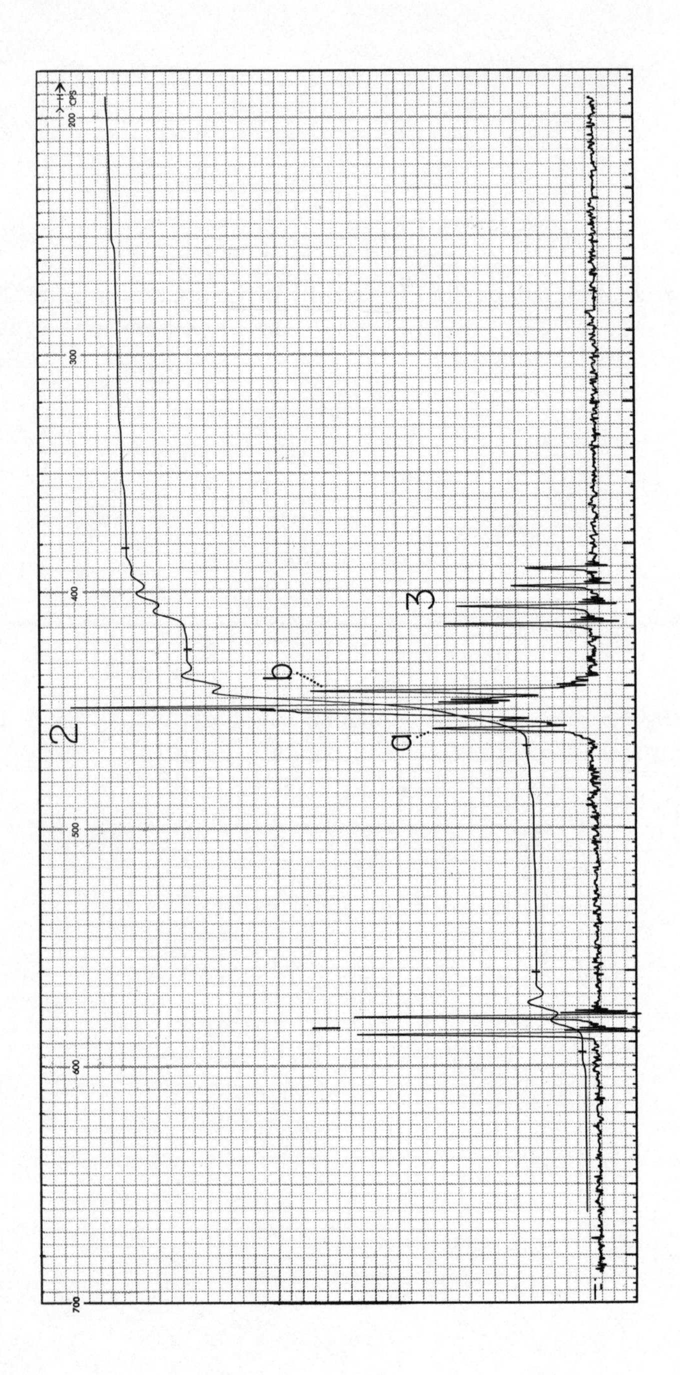

200 CPS
300
400
500
600
700
1
2
3
a
b

This is the spectrum of one of the following two isomers:

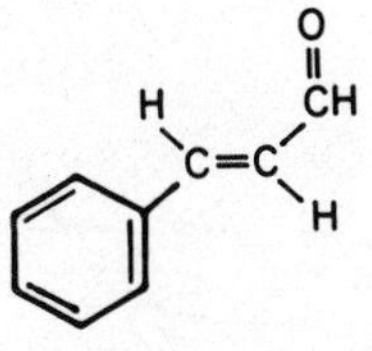

I 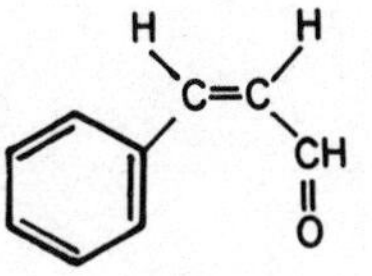    II

$C_9H_8O$

The scan over the 700 to 200-cps region is given.

| Band | ① | ② | ③ |
|---|---|---|---|
| Integration | 19 | 143 | 25 |

---

*Questions*

1. What spin systems are involved here?
2. Analyze the three-spin pattern on the assumption that it is first-order.
3. On the basis of this analysis, which is the correct structure, I or II?
4. If this pattern is analyzed as an ABX, where are the eight peaks expected in the AB portion and the six peaks expected in the X portion?

Answers: p. 235.

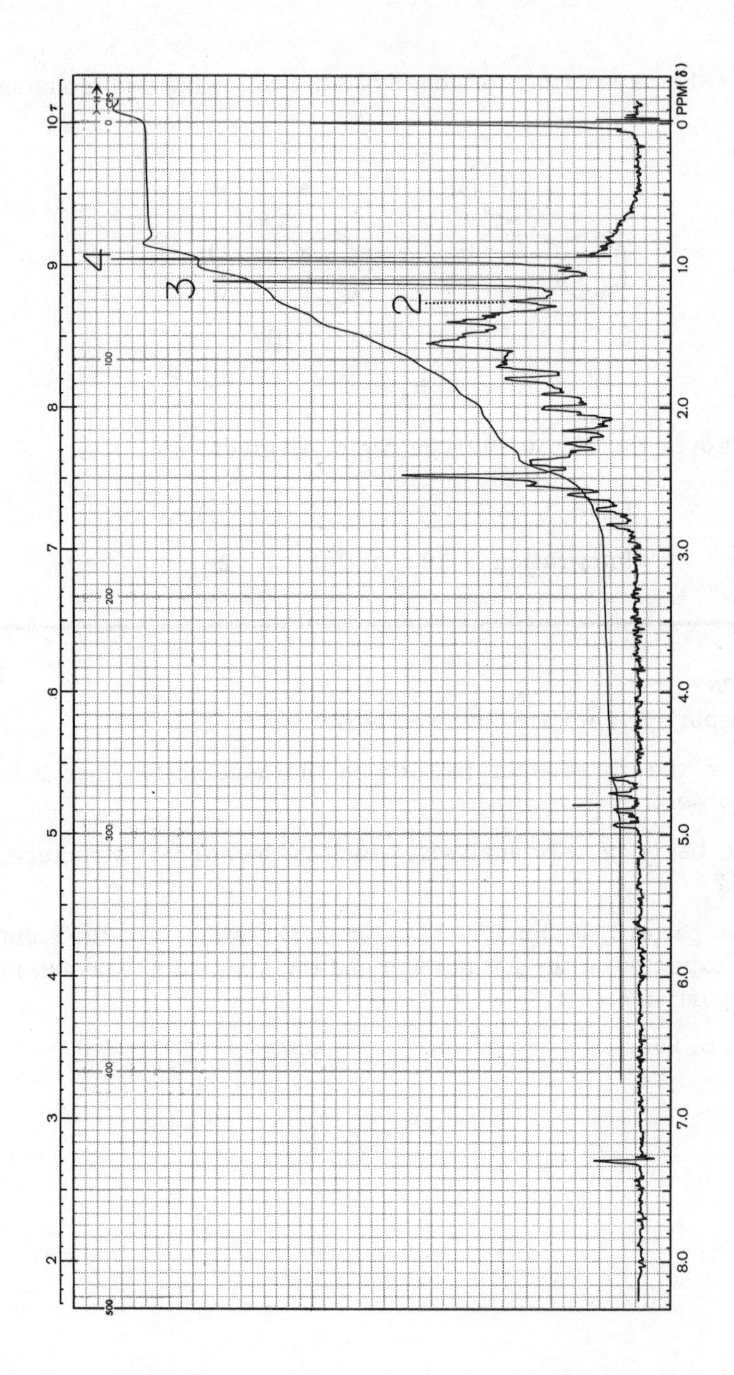
0 PPM(δ)
1.0
2.0
3.0
4.0
5.0
6.0
7.0
8.0
10 τ
9
8
7
6
5
4
3
2
CPS
0
100
200
300
400
500
4
3
2
1

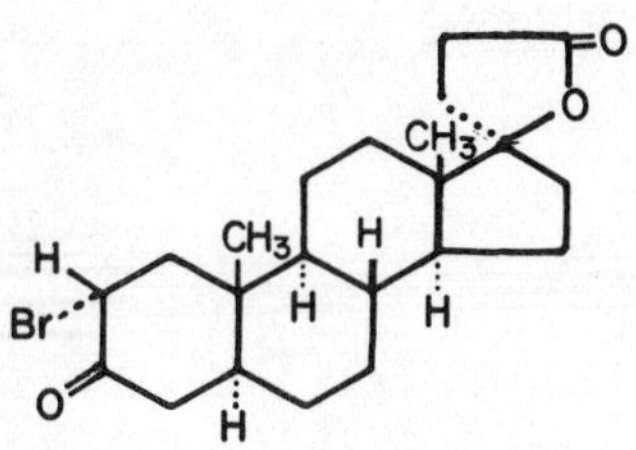

$C_{22}H_{31}BrO_3$

Band ① integrates for 4 units, while the entire spectrum integrates as 199 units.

---

*Questions*

1. Is this compound pure?
2. In notational terms, what does the pattern in band ① represent?
3. Can the splittings seen in band ① be taken as coupling constants?
4. What can be deduced from the pattern in band ①?

Answers: p. 235.

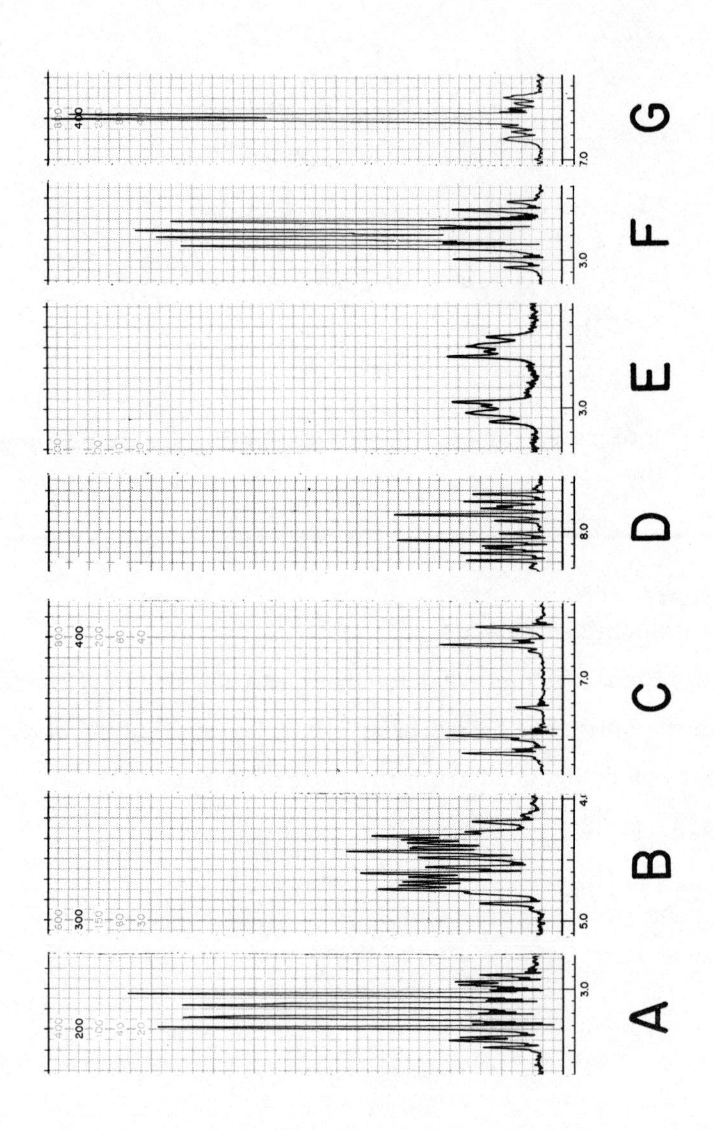
A
B
C
D
E
F
G

Each of the spin patterns at the left is due to four or eight protons. All of these patterns are reproduced from the Varian Catalogs by permission of Varian Associates.

---

*Questions*

1. What common characteristic do these patterns have?
2. Choose the pattern or patterns which correspond to the following systems: (a) *p*-disubstituted phenyl; (b) aliphatic $-CH_2-CH_2-$ group; (c) flexible six-membered ring containing two $-CH_2-CH_2-$ groups; and (d) symmetrical *o*-disubstituted phenyl.
3. Describe these spin patterns in notational terms.

Answers: p. 236.

# SECTION 7

General problems involving higher-order patterns.

*Suggested Reading*

Bible: pp. 105-111.
Emsley, Feeney, and Sutcliffe: pp. 240-244.
Jackman: pp. 82-83.
Roberts: pp. 86-87.

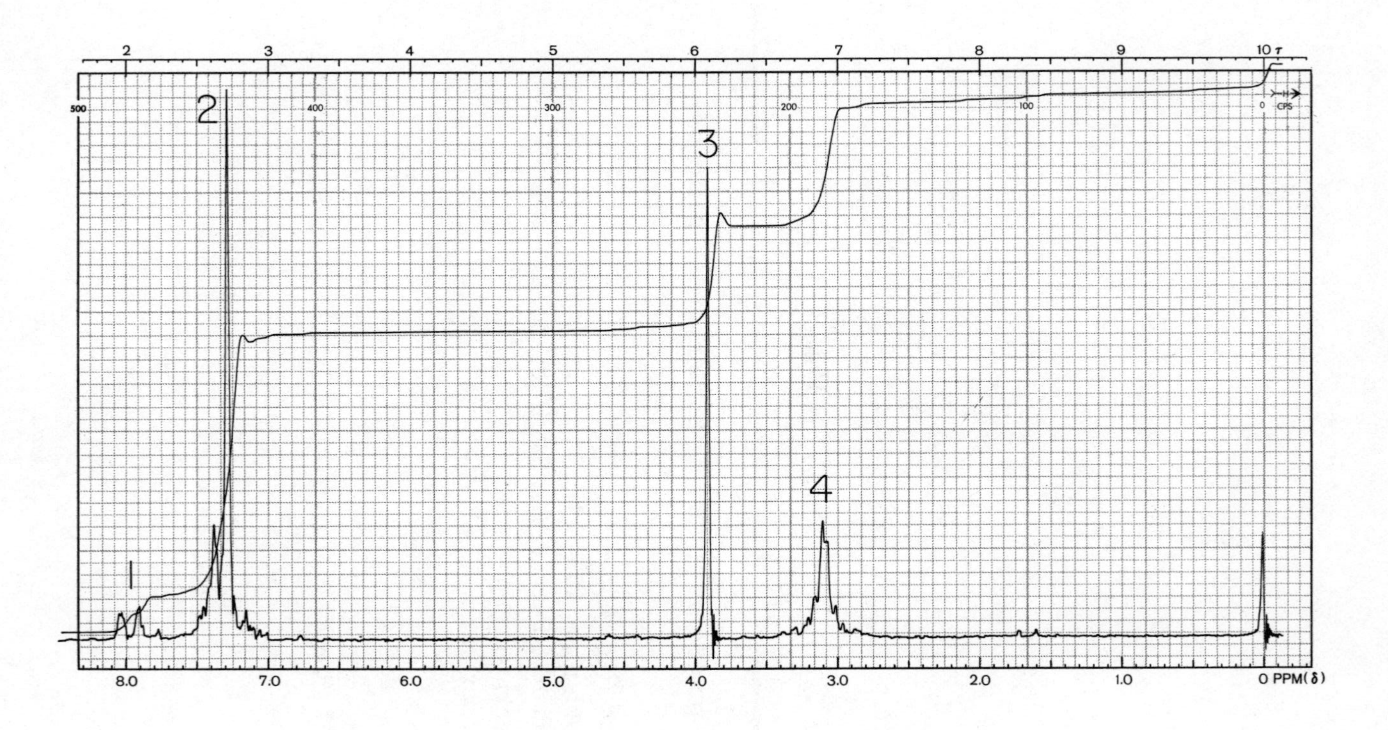

2
3
4
5
6
7
8
9
10 τ
500
400
300
200
100
0
CPS
1
2
3
4
8.0
7.0
6.0
5.0
4.0
3.0
2.0
1.0
0 PPM(δ)

$C_{16}H_{16}O_2S$

| Band | ① | ② | ③ | ④ | Sum |
|---|---|---|---|---|---|
| Integration | 15 | 109 | 43 | 52 | 219 |

None of these signals are removed by $D_2O$ exchange. There are no signals in the 1000 – 500-cps region.

---

*Questions*

1. Propose a structure consistent with this spectrum.
2. Describe all of the spin systems in notational form.

Answers: p. 238.

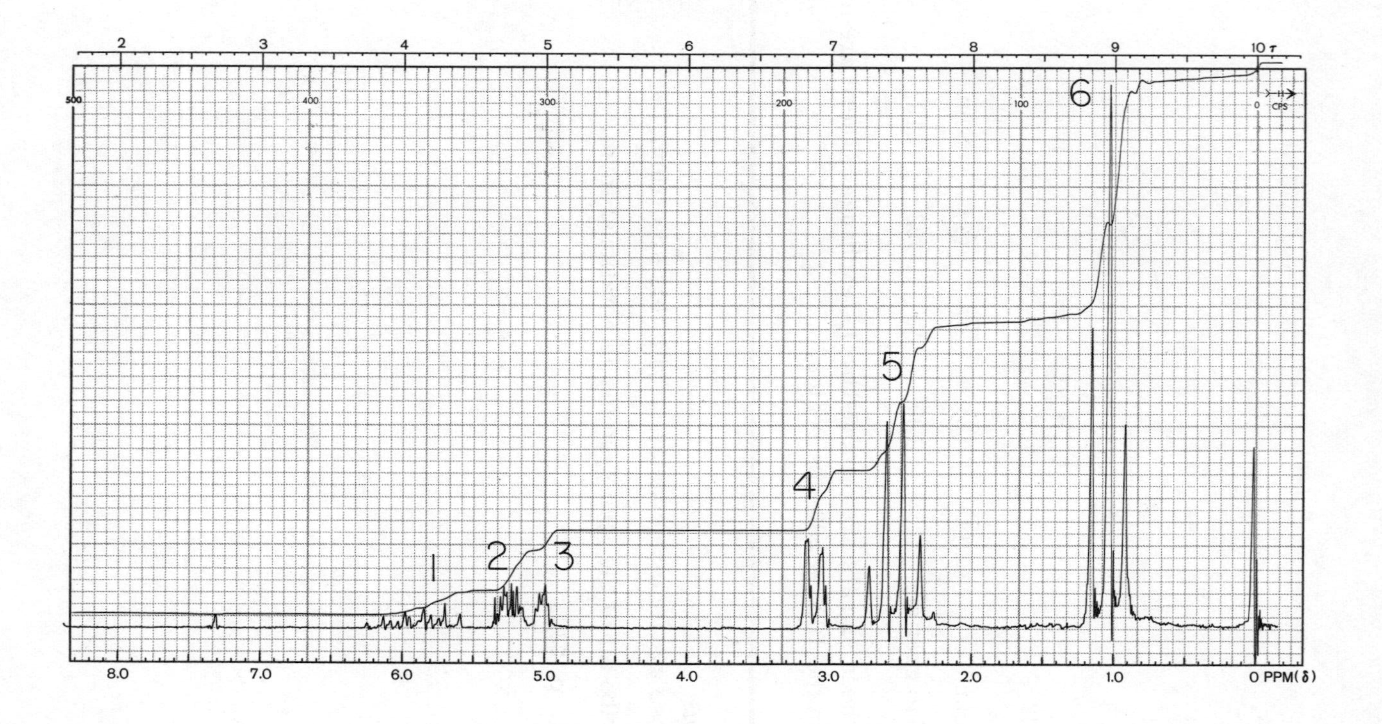
2
3
4
5
6
7
8
9
10 τ
500
400
300
200
100
0
CPS
1
2
3
4
5
6
8.0
7.0
6.0
5.0
4.0
3.0
2.0
1.0
0 PPM(δ)

$$\begin{array}{c} H \quad\quad H \\ \diagdown C=C \diagup \\ \diagup \quad\quad \diagdown \\ H \quad\quad CH_2N(CH_2CH_3)_2 \end{array}$$

$C_7H_{15}N$

| Band | ① | ② | ③ | ④ | ⑤ | ⑥ |
|---|---|---|---|---|---|---|
| Integration | 10 | 16.5 | 8 | 25 | 60 | 100 |

---

*Questions*

1. Assign the numbered bands.
2. Give a notational description of the spin systems involved.
3. Give a qualitative description of the splittings.

Answers: p. 239.

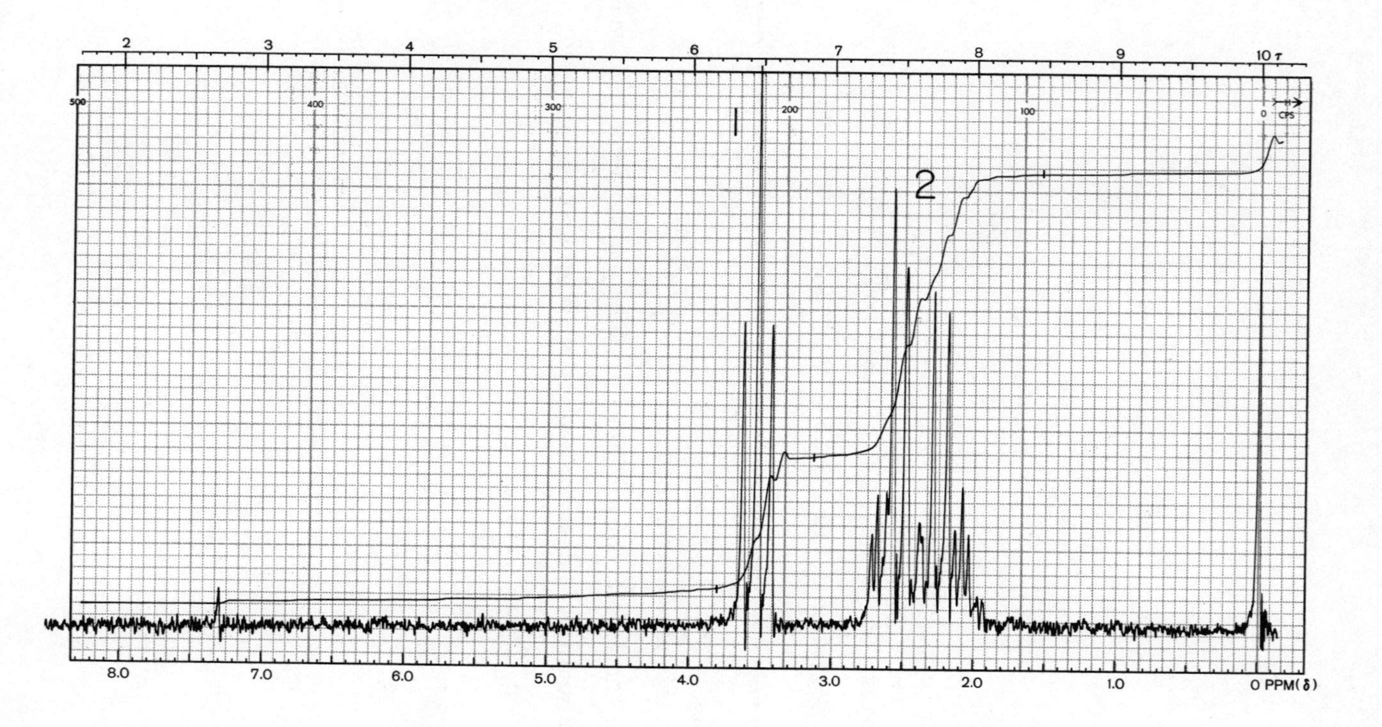
2
3
4
5
6
7
8
9
10 τ
500
400
300
200
100
0
CPS
1
2
8.0
7.0
6.0
5.0
4.0
3.0
2.0
1.0
0 PPM(δ)

$C_4H_6BrN$

None of the signals are removed by $D_2O$ exchange.

| Band | ① | ② |
|---|---|---|
| Integration | 55 | 124 |

---

*Questions*

1. Propose a structure for this compound which is consistent with the spectrum.
2. Describe the spin system or systems in notational form.

Answers: p. 240.

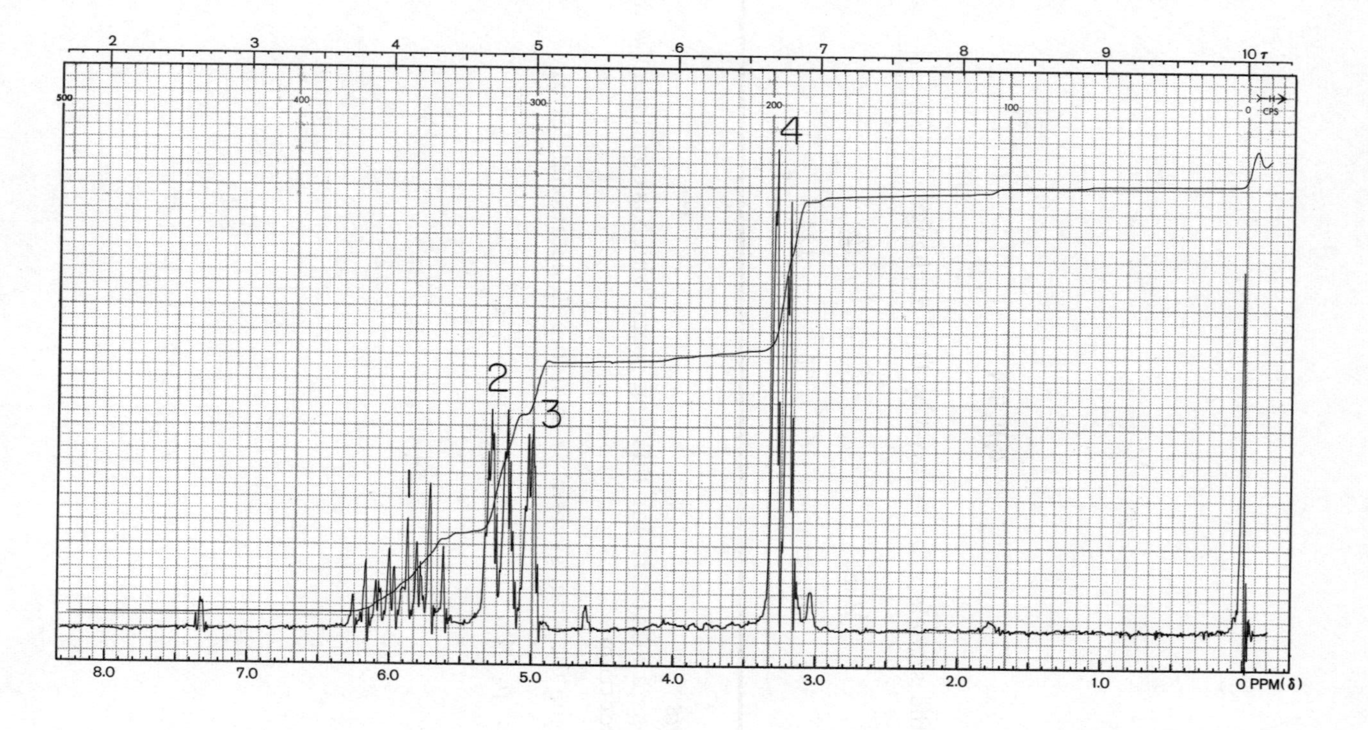
2
3
4
5
6
7
8
9
10 τ
500
400
300
200
100
0
CPS
4
2
3
8.0
7.0
6.0
5.0
4.0
3.0
2.0
1.0
0 PPM(δ)

A $CDCl_3$ solution of a compound having the molecular formula $C_6H_{11}N$ was treated with $D_2O$. This treatment removed a sharp one-proton signal at 86 cps. The $D_2O$-exchange spectrum is shown at the left.

| Band | ① | ② + ③ | ④ |
|---|---|---|---|
| Integration | 34 | 72 | 64 |

---

*Questions*

1. Propose a structure for the compound which is consistent with the spectrum.
2. How could the extraction of the coupling constants and chemical shifts be made much easier?

Answers: p. 240.

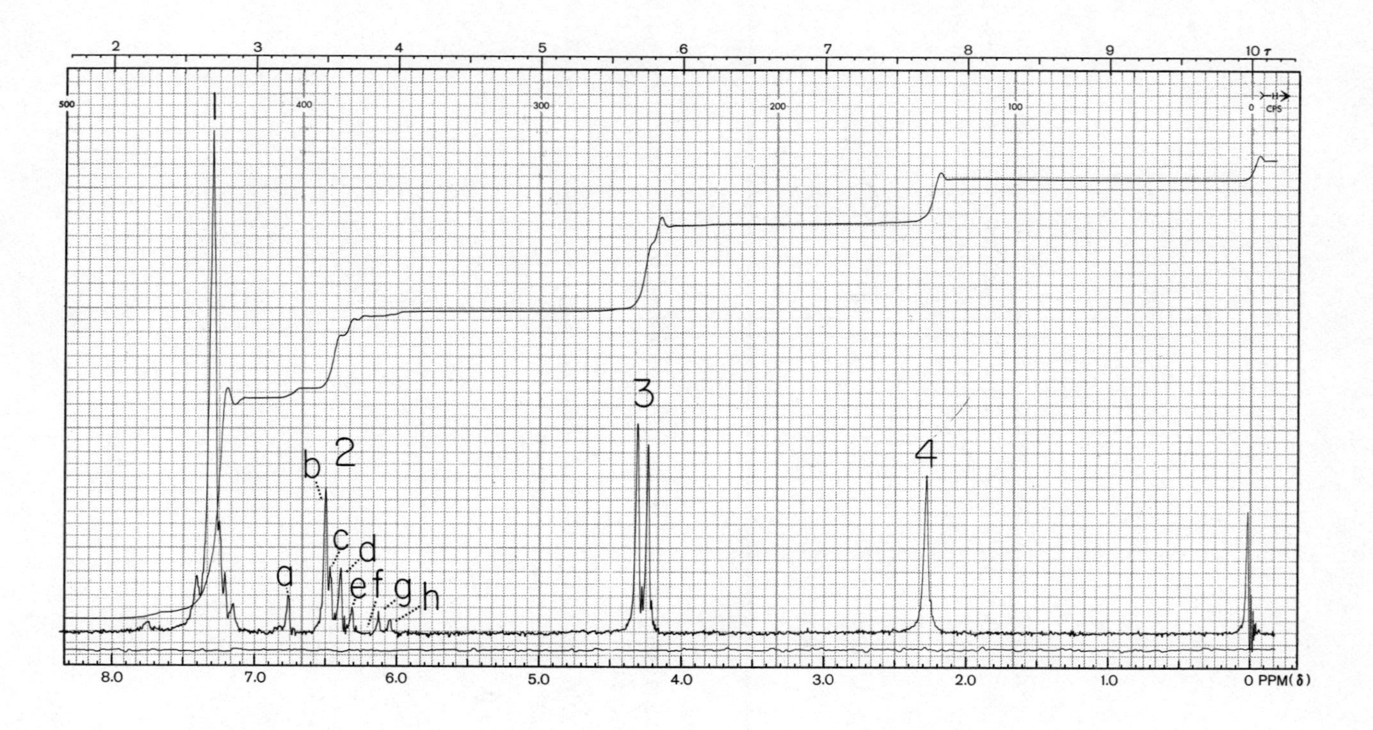

2
3
4
5
6
7
8
9
10 τ
500
400
300
200
100
0
CPS
1
2
3
4
a
b
c
d
e
f
g
h
8.0
7.0
6.0
5.0
4.0
3.0
2.0
1.0
0 PPM(δ)

$C_9H_{10}O$

The lower trace is the spectrum over the region 1000 – 500 cps. The only change which takes place on $D_2O$ exchange is the loss of band ④.

| Band | ① | ② | ③ | ④ |
|---|---|---|---|---|
| Integration | 92 | 36 | 35.5 | 18 |

---

*Questions*

1. Propose a structure for the compound which is consistent with the spectrum.
2. In notational terms, what is the pattern in bands ② and ③?

Answers: p. 241.

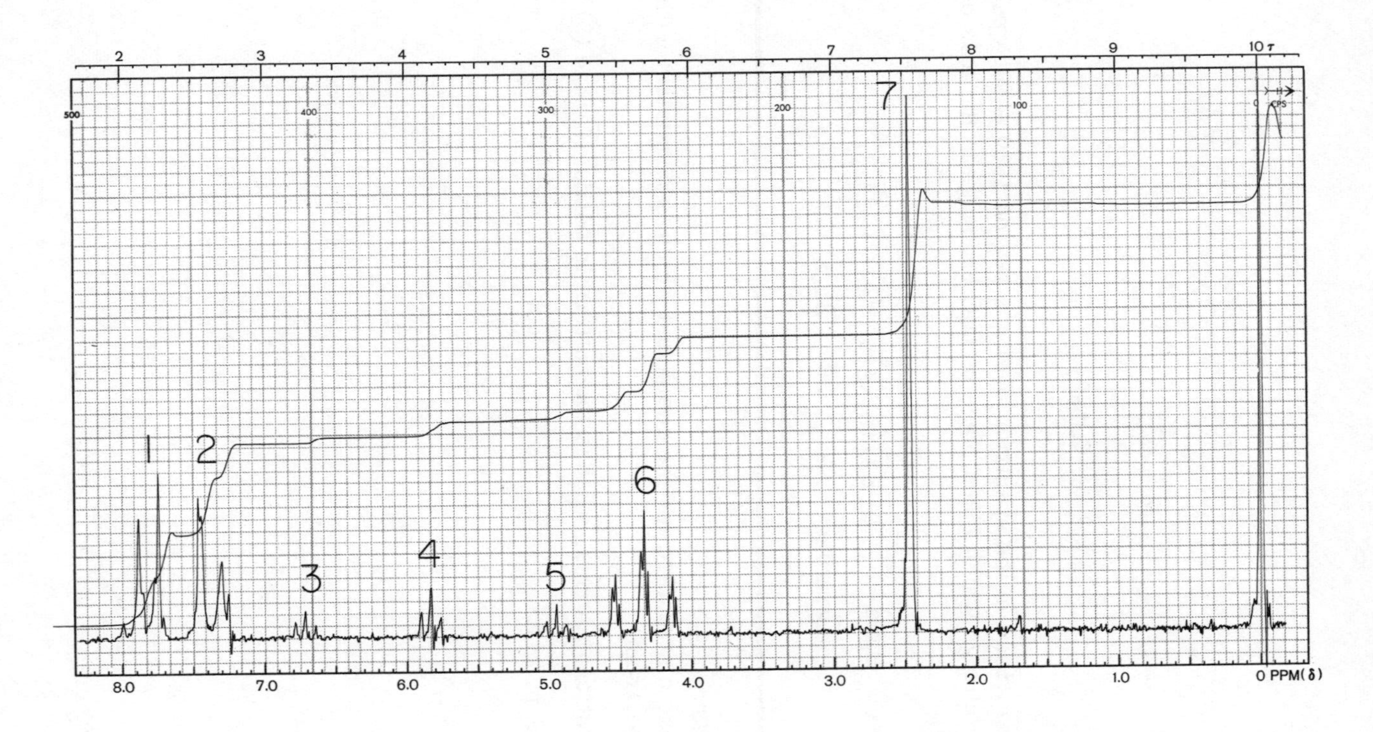
1
2
3
4
5
6
7
0 PPM(δ)
10 τ
CPS

$H-CF_2-CF_2-CH_2-O-SO_2-C_6H_4-CH_3$

$C_{10}H_{10}O_3F_4S$

| Band | ① | ② | ③ | ④ | ⑤ | ⑥ | ⑦ |
|---|---|---|---|---|---|---|---|
| Integration | 37 | 38 | 2 | 6 | 3 | 31 | 55 |

*Questions*

1. Assign all of the bands to specific protons.
2. Explain why the peaks in band ② are broader than those in band ①.
3. Which of the proton–fluorine coupling constants can be measured with assurance from this spectrum?
4. Is the distortion in intensities seen in the closely spaced triplets of band ⑥ probably due to departure of the system from first-order rules? What is the probable cause?
5. How could the explanation of the distortion seen in the closely spaced triplets of band ⑥ be checked?

Answers: p. 241.

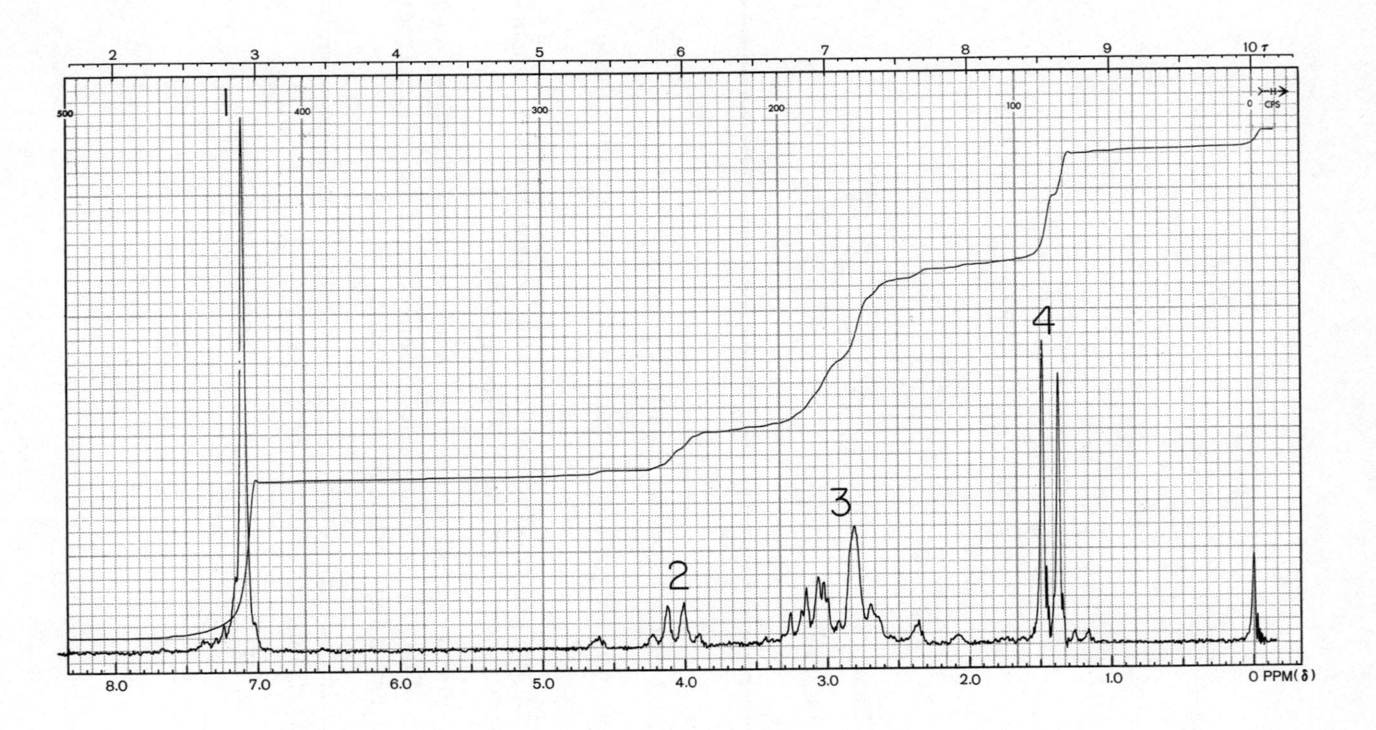

2
3
4
5
6
7
8
9
10 τ
500
400
300
200
100
0 CPS
1
2
3
4
8.0
7.0
6.0
5.0
4.0
3.0
2.0
1.0
0 PPM(δ)

This spectrum and the spectra on pp. 130 and 132 were determined using, in each case, one of the following compounds:

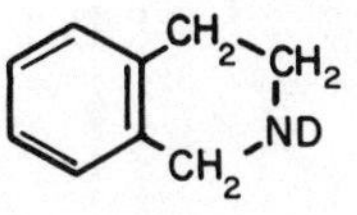

| I | II | III |
|---|---|---|
| $C_9H_{10}DN$ | $C_{10}H_{12}DN$ | $C_{10}H_{12}DN$ |

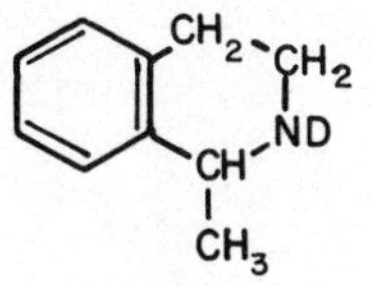

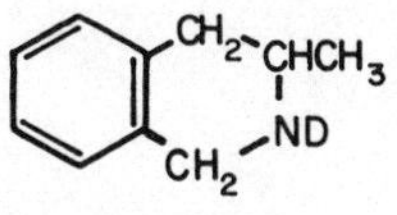

Before $D_2O$ exchange, each of the compounds showed an —NH— proton signal between 93 and 109 cps.

---

*Questions*

1. Match each spectrum with the corresponding compound.
2. Can any of the separations in band ③ be taken as coupling constants?

Answers: p. 243.

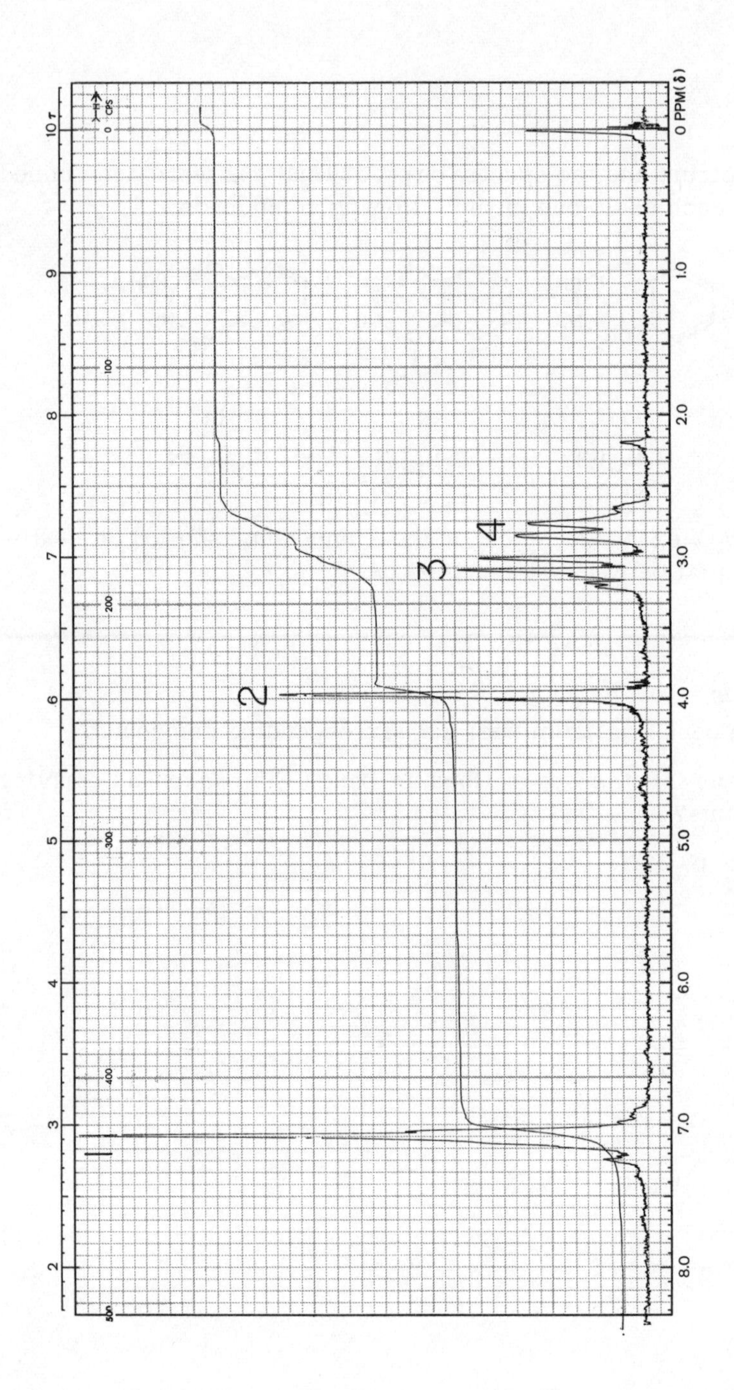

This spectrum was introduced on p. 129. From the molecular structure, one might suspect that bands ③ and ④ should represent an $A'_2B'_2$ system.

---

*Questions*

1. Cite possible reasons for the lack of symmetry about the center of the $A'_2B'_2$ pattern.

2. How could one choose the correct reason for the lack of symmetry about the center of the $A'_2B'_2$ pattern?

Answers: p. 243.

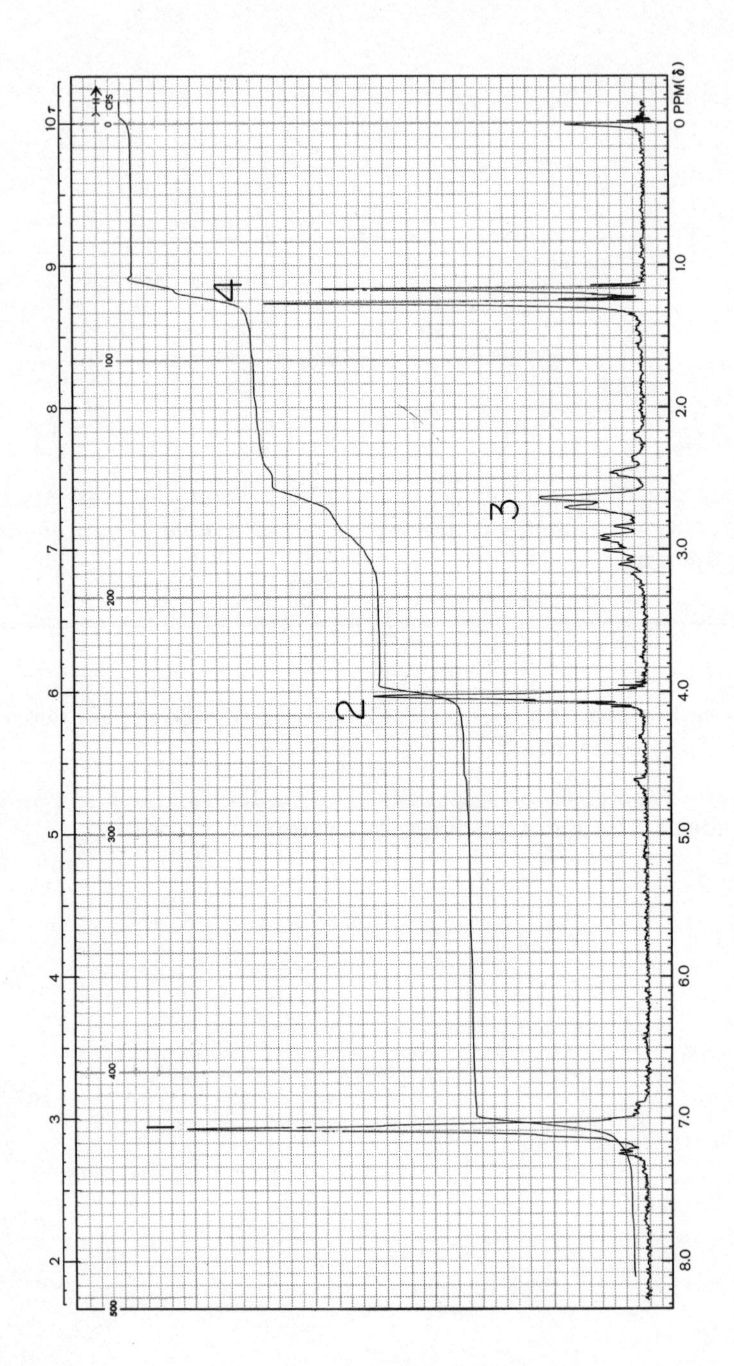
10 τ
9
8
7
6
5
4
3
2
0
100
200
300
400
500
CPS
4
3
2
0 PPM (δ)
1.0
2.0
3.0
4.0
5.0
6.0
7.0
8.0

This spectrum was introduced on p. 129.

---

*Questions*

1. Suggest methods by which band ③ might be simplified in order to extract more information.

2. In the spectrum on p. 128, there are two small peaks at a higher field than band ④. Suggest a possible explanation for these small peaks.

Answers: p. 244.

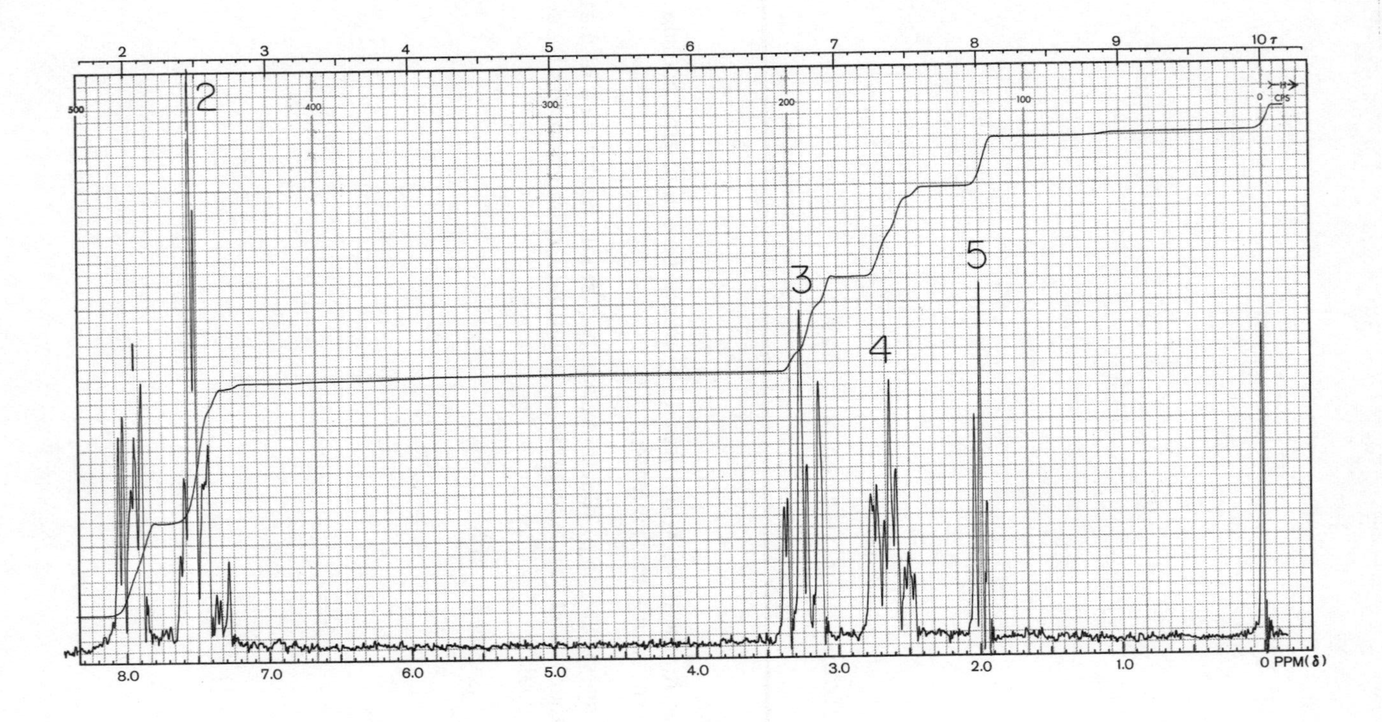
2
3
4
5
6
7
8
9
10 τ
500
400
300
200
100
0 CPS
8.0
7.0
6.0
5.0
4.0
3.0
2.0
1.0
0 PPM (δ)
1
2
3
4
5

$C_{11}H_{10}O$

None of the signals are removed by $D_2O$ exchange.

| Band | ① | ② | ③ | ④ | ⑤ | Sum |
|---|---|---|---|---|---|---|
| Integration | 38.5 | 59 | 39.5 | 37.5 | 20.5 | 195.0 |

---

*Questions*

1. Propose a structure for this compound.
2. Describe the spin systems in notational terms.
3. Explain the splittings.
4. In the absence of coupling between the protons represented by bands ④ and ⑤, what would be the appearance of band ④?

Answers: p. 244.

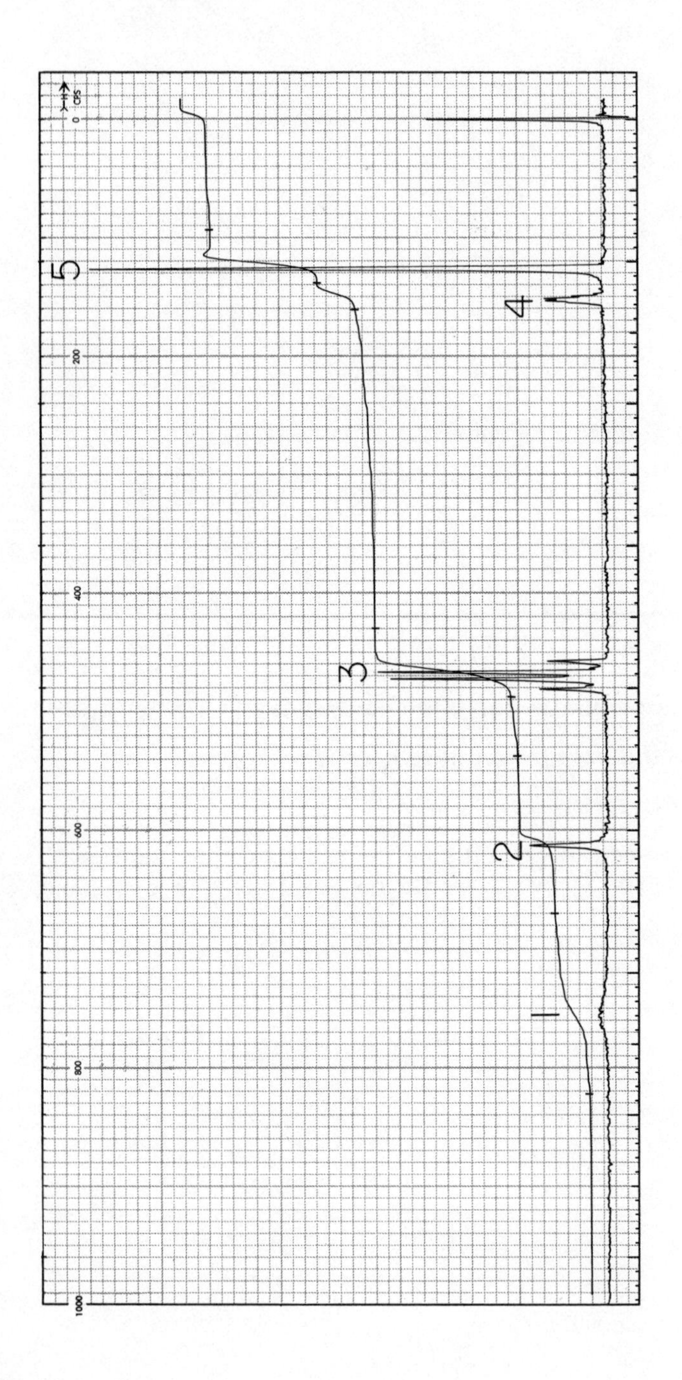

0
200
400
600
800
1000
CPS
1
2
3
4
5

$C_9H_9NO_3$

This spectrum was determined in $(CD_3)_2SO$ over the region 1000 – 0 cps. The spectrum on p. 138 was determined after the solution had been diluted with a few drops of $D_2O$.

| Band | ① | ② | ③ | ④ | ⑤ |
|---|---|---|---|---|---|
| Integration | 14 | 14.5 | 57 | 17 | 45.5 |

---

*Questions*

1. Point out the absorption band which is due to the solvent. Explain any multiplicity which should be expected in this band.
2. Propose a structure for the molecule.

Answers: p. 246.

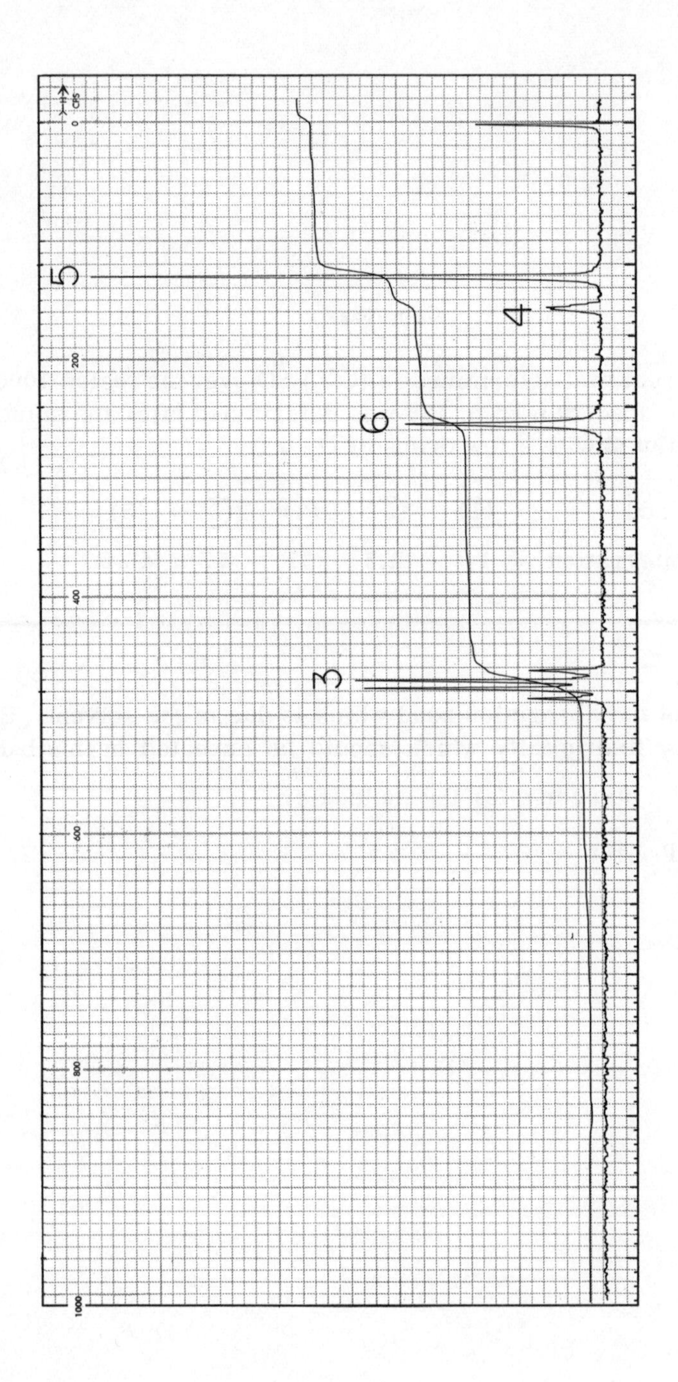
0 CPS
200
400
600
800
1000
5
4
6
3

The solution used for the spectrum on p. 136 was diluted with several drops of $D_2O$. The spectrum was then rerun.

---

*Questions*

1. Explain the changes that took place on dilution.

2. What can be said about the rate of exchange between the $-\overset{O}{\overset{\|}{C}}NH-$ and $-COOH$ protons in the solution used for the spectrum on p. 136?

Answers: p. 248.

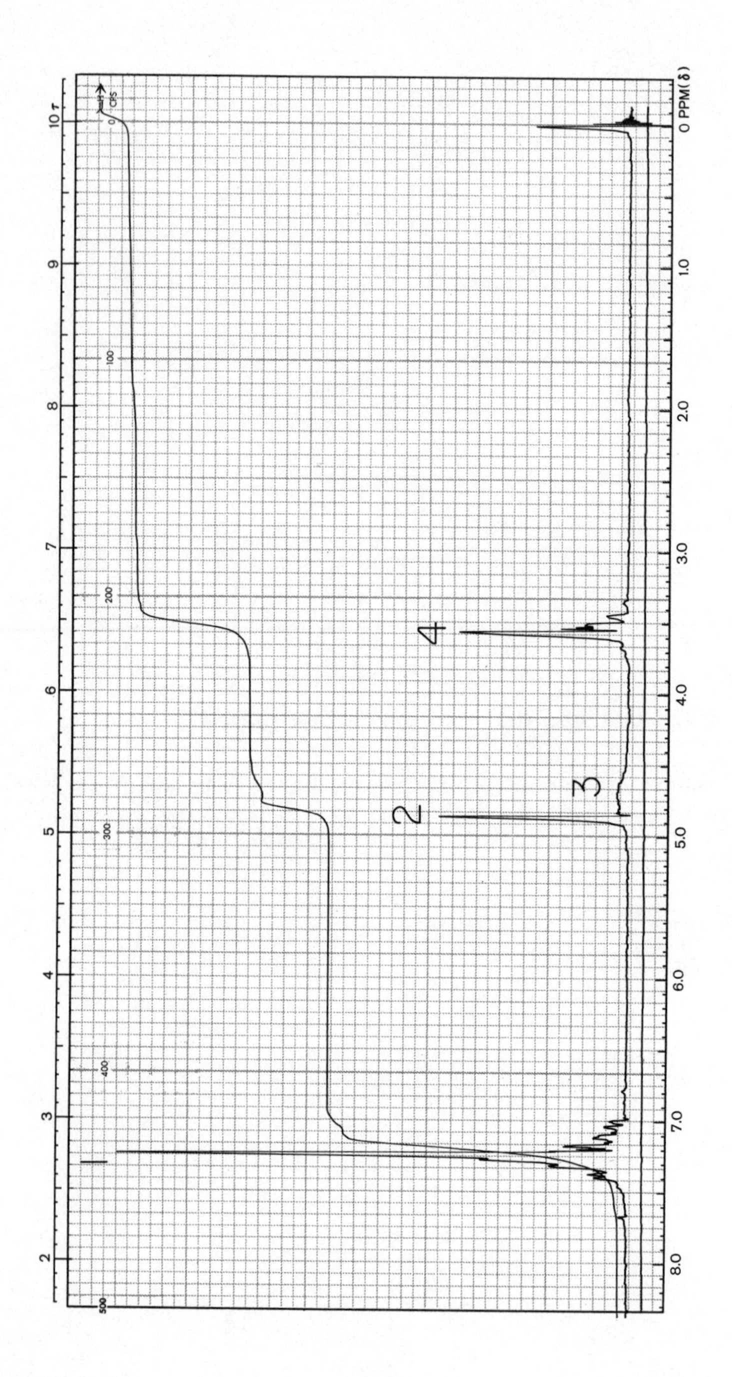

10 τ
9
8
7
6
5
4
3
2
0 PPM (δ)
1.0
2.0
3.0
4.0
5.0
6.0
7.0
8.0
CPS
0
100
200
300
400
500
1
2
3
4

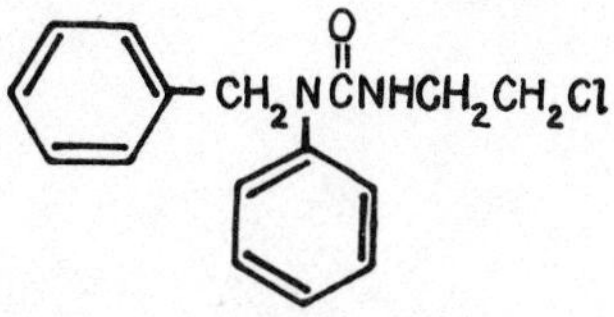

$\overset{O}{\parallel}$

$CH_2N\overset{}{C}NHCH_2CH_2Cl$

$C_{16}H_{17}ClN_2O$

| Band | ① | ② + ③ | ④ |
|---|---|---|---|
| Integration | 121 | 33.5 | 47 |

The lowest trace is the spectrum from 1000 to 500 cps.

---

*Questions*

1. Assign the bands to specific protons.
2. Explain the appearance of band ④.

Answers: p. 248.

# SECTION 8

Unexpected nonequivalence of protons involving higher-order spin systems.

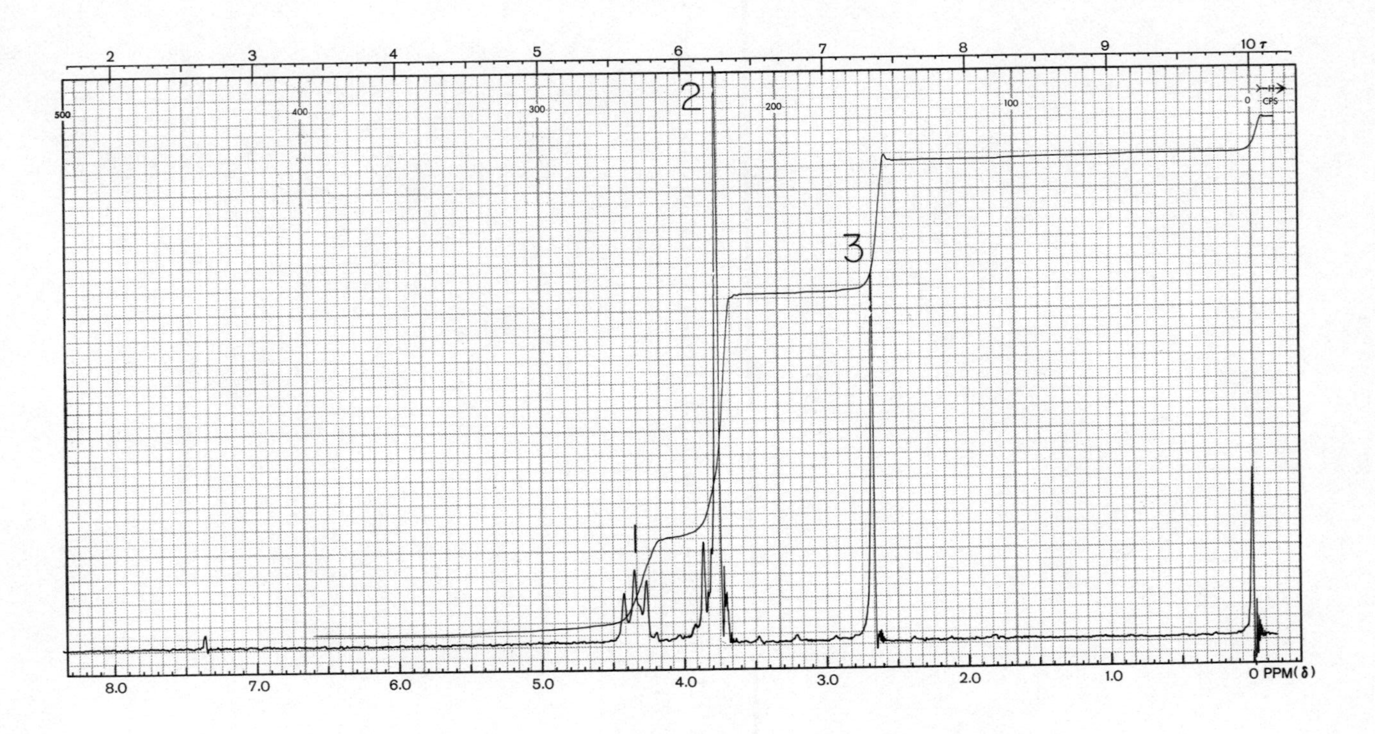

2
3
4
5
6
7
8
9
10 τ
500
400
300
200
100
0 CPS
2
3
1
8.0
7.0
6.0
5.0
4.0
3.0
2.0
1.0
0 PPM(δ)

$$CH_3\overset{S}{\overset{\|}{C}}N\langle^{CH_2-CH_2}_{CH_2-CH_2}\rangle O$$

$C_6H_{11}NOS$

| Band | ① | ② | ③ | Sum |
|---|---|---|---|---|
| Integration | 39 | 101 | 54 | 194 |

---

*Questions*

1. Assign the bands and explain the appearance of bands ① and ②.
2. What change would probably be observed in the spectrum if the temperature of the solution were increased?

Answers: p. 249.

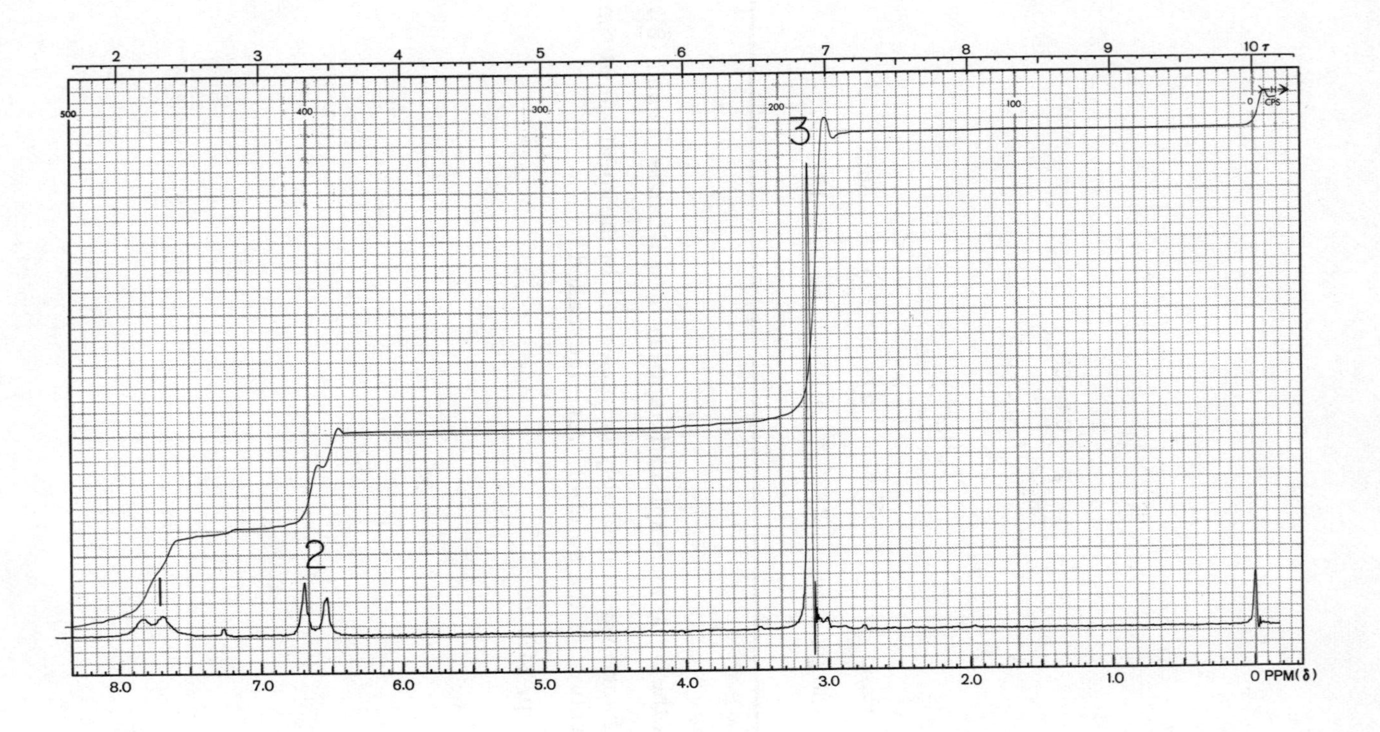
2
3
4
5
6
7
8
9
10 τ
500
400
300
200
100
0
CPS
1
2
3
8.0
7.0
6.0
5.0
4.0
3.0
2.0
1.0
0 PPM(δ)

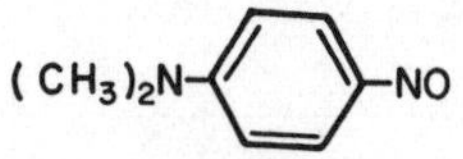

$(CH_3)_2N-C_6H_4-NO$

$C_8H_{10}N_2O$[1]

| Band | ① | ② | ③ |
|---|---|---|---|
| Integration | 38 | 40 | 122 |

---

*Questions*

1. Assign the numbered bands.
2. Why is band ① broader than band ②?

Answers: p. 250.

[1] The spectrum of this compound has been studied in detail by D. D. MacNicol, R. Wallace, and J. C. D. Brand, Trans. Faraday Soc. 61:1 (1965) and I. R. King and G. W. Kirby, J. Chem. Soc. 1334 (1966).

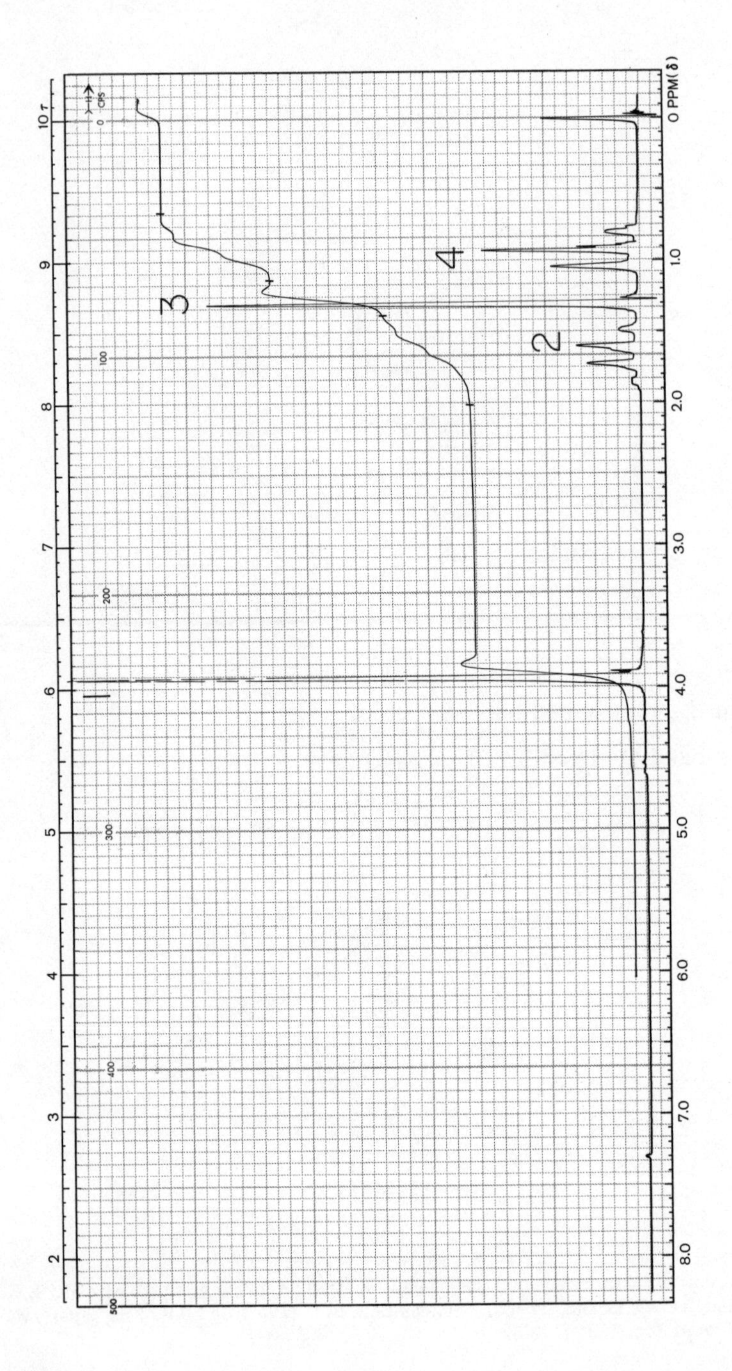

10 τ
9
8
7
6
5
4
3
2
0
100
200
300
400
500
CPS
0 PPM(δ)
1.0
2.0
3.0
4.0
5.0
6.0
7.0
8.0
3
4
2
1

$$\begin{array}{c} CH_2{-}CH_2 \\ | \quad\quad | \\ O \quad\; O \\ \diagdown \; \diagup \\ CH_3{-}C{-}CH_2CH_3 \end{array}$$

$C_6H_{12}O_2$

| Band | ① | ② | ③ | ④ |
|---|---|---|---|---|
| Integration | 64 | 36 | 47 | 45 |

---

*Questions*

1. Assign all of the bands to specific groups of protons.
2. Describe all of the groups of protons in notational terms.
3. Explain the splitting observed in bands ② and ④.
4. Why do bands ① and ③ appear as single sharp peaks rather than as multiplets?

Answers: p. 250.

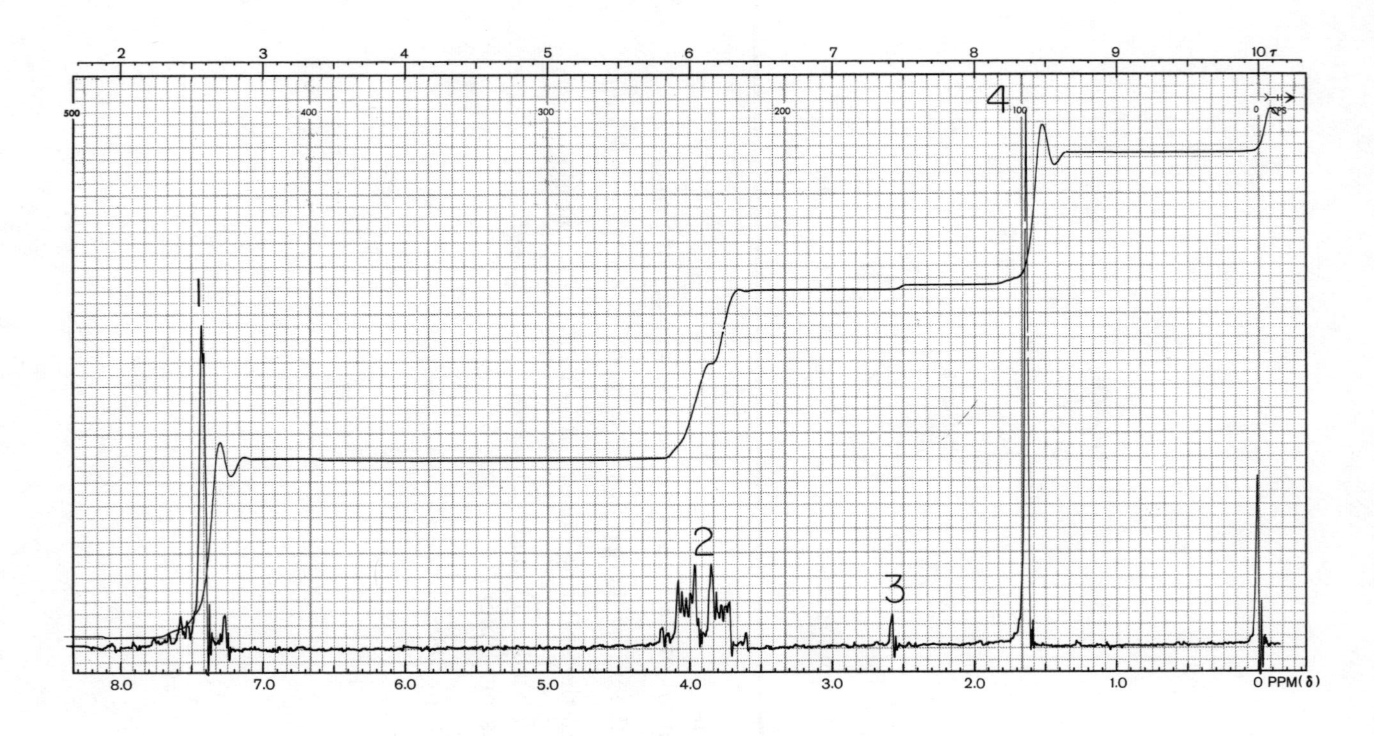

2
3
4
5
6
7
8
9
10 τ
500
400
300
200
100
0 CPS
1
2
3
4
8.0
7.0
6.0
5.0
4.0
3.0
2.0
1.0
0 PPM(δ)

This is the spectrum of the crude product of the following reaction:

$$\text{Br}-\text{C}_6\text{H}_4-\overset{\text{O}}{\overset{\|}{\text{C}}}-\text{CH}_3 \longrightarrow \text{Br}-\text{C}_6\text{H}_4-\text{C}(\text{O}-\text{CH}_2-\text{CH}_2-\text{O})-\text{CH}_3$$

$C_8H_7BrO$ $\qquad$ $C_{10}H_{11}BrO_2$

| Band | ① | ② | ③ | ④ |
|---|---|---|---|---|
| Integration | 74 | 70 | 1.8 | 54.5 |

---

*Questions*

1. Assign all of the bands.
2. Identify the pattern in band ② and explain its origin. Why is this pattern different from the pattern seen in the spectrum on p. 148?
3. To what extent did the reaction proceed?

Answers: p. 251.

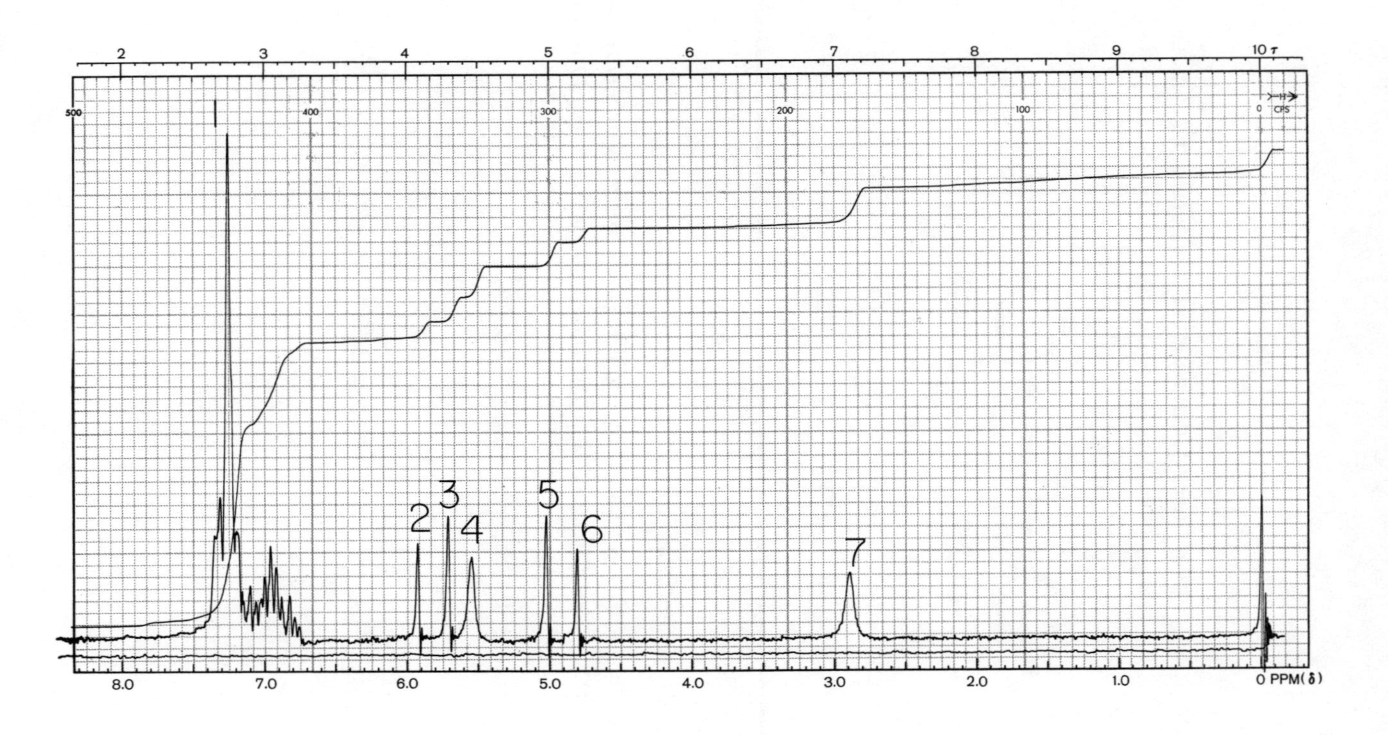
2
3
4
5
6
7
8
9
10 τ
500
400
300
200
100
0 CPS
1
2
3
4
5
6
7
8.0
7.0
6.0
5.0
4.0
3.0
2.0
1.0
0 PPM(δ)

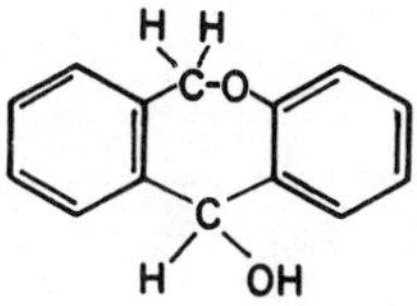

$C_{14}H_{12}O_2$

The spectrum after $D_2O$ exchange is shown on p. 154.

| Band | ① | ② | ③ | ④ | ⑤ | ⑥ | ⑦ |
|---|---|---|---|---|---|---|---|
| Integration | 117 | 6.5 | 10 | 13 | 10 | 5.5 | 14 |

---

*Questions*

1. Assign band ① and peaks ② – ⑦.
2. Give a reason for the broadness of peaks ④ and ⑦.
3. Suggest a way of causing peaks ④ and ⑦ to become sharper.

Answers: p. 252.

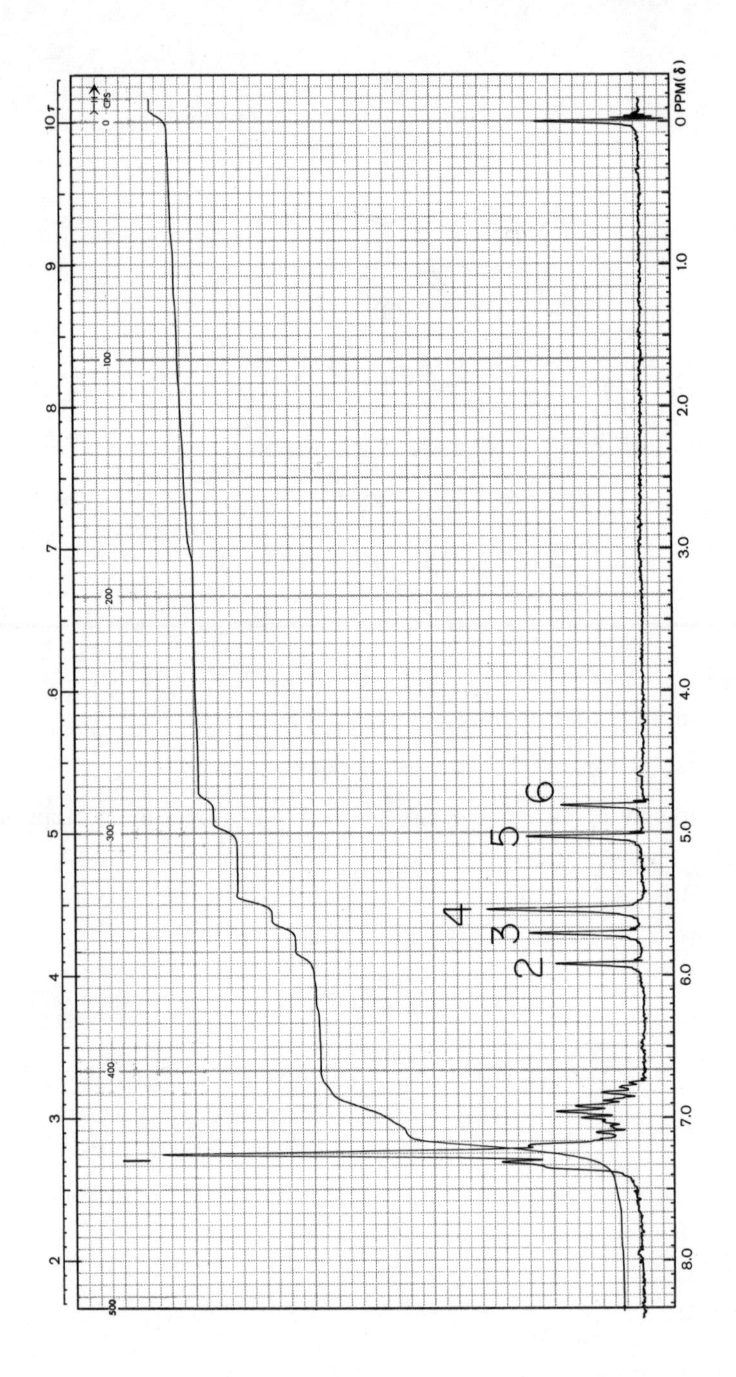

1
2
3
4
5
6
0 PPM(δ)
1.0
2.0
3.0
4.0
5.0
6.0
7.0
8.0
10τ
9
8
7
6
5
4
3
2
0
100
200
300
400
500
CPS

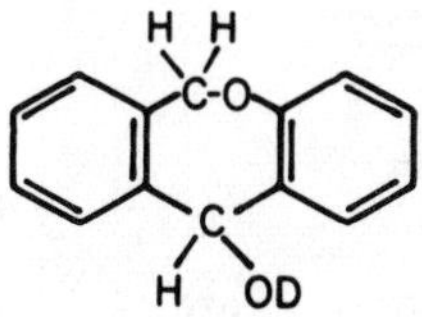

$C_{14}H_{11}DO_2$

The spectrum before $D_2O$ exchange is shown on p. 152.

---

*Question*

1. Suggest changes which might be brought about in the spectrum by preparation of the acetate.

Answer: p. 252.

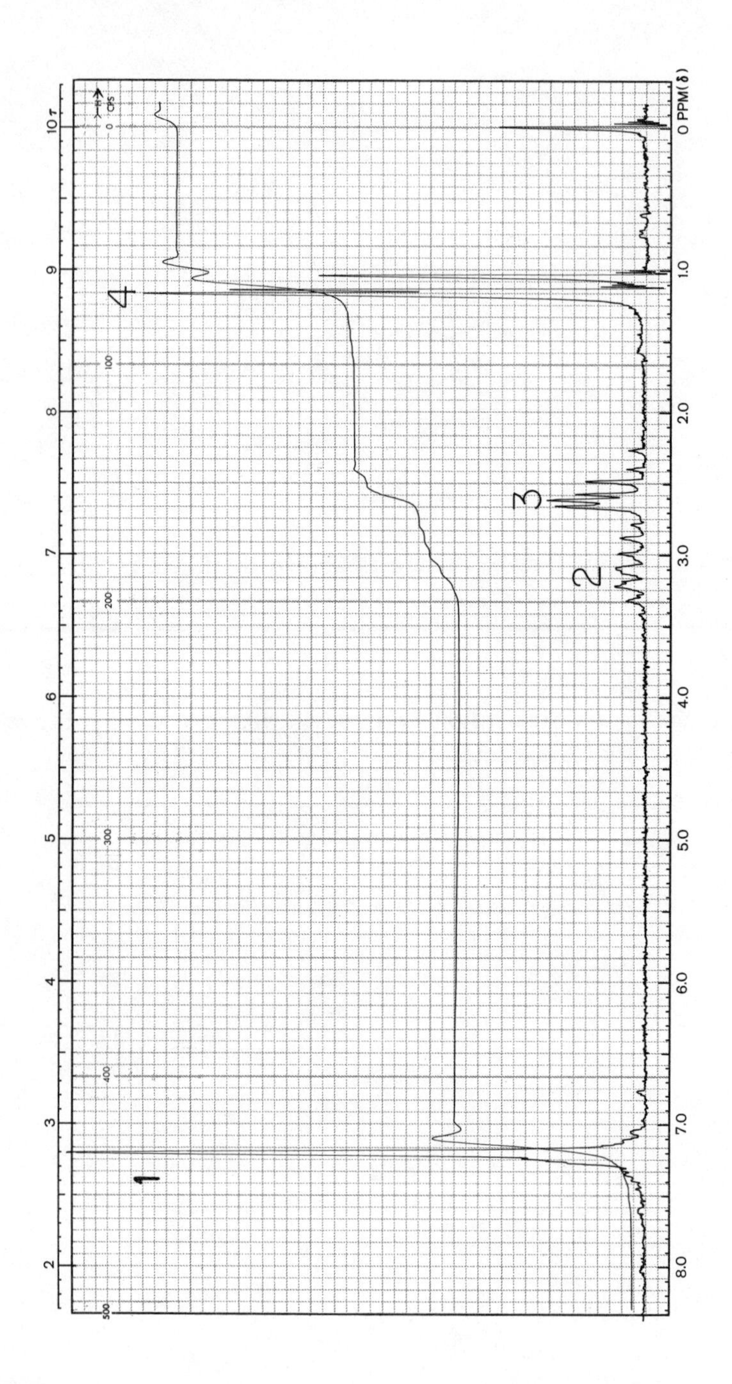

This spectrum and the one on p. 158 were determined on two different occasions using the following compound:

$C_6H_5$–$CH_2CH(NH_2)CH_3$

$C_9H_{13}N$

| Band | ① | ② | ③ | ④ | Sum |
|---|---|---|---|---|---|
| Integration | 73.5 | 14 | 28.5 | 73.5 | 189.5 |

---

*Questions*

1. Assign all of the signals to specific protons.
2. Explain why band ④ in the spectrum on the left is different from band ④ in the spectrum on p. 158.

Answers: p. 253.

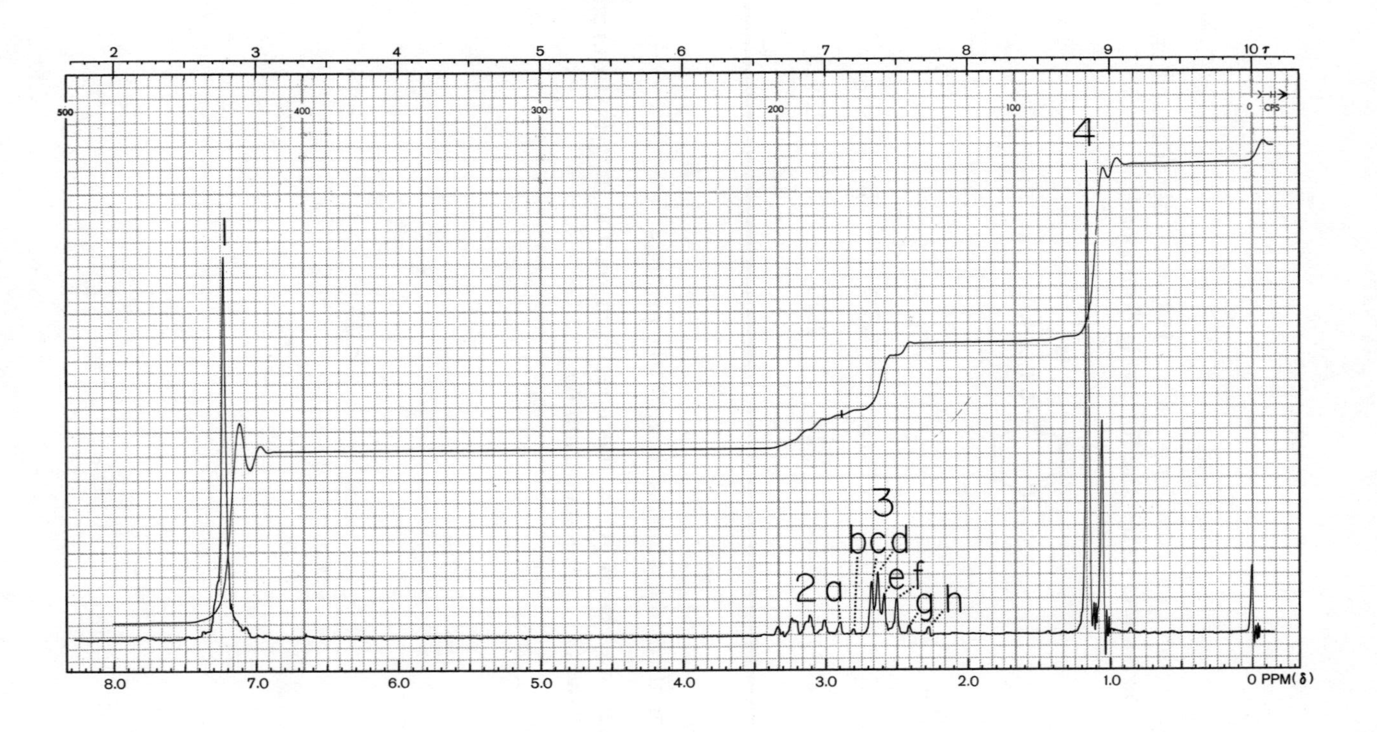

10 τ
CPS
4
1
3
bcd
2 a
e f
g h
0 PPM(δ)

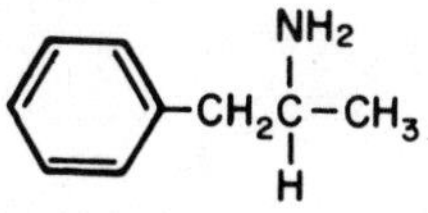

$C_9H_{13}N$

This spectrum was introduced on p. 157.

| Band | ① | ② | ③ | ④ |
| --- | --- | --- | --- | --- |
| Integration | 71 | 13.5 | 29 | 74 |

---

*Questions*

1. Explain the complexity of band ③.
2. Which of the coupling constants can be determined by inspection?

Answers: p. 253.

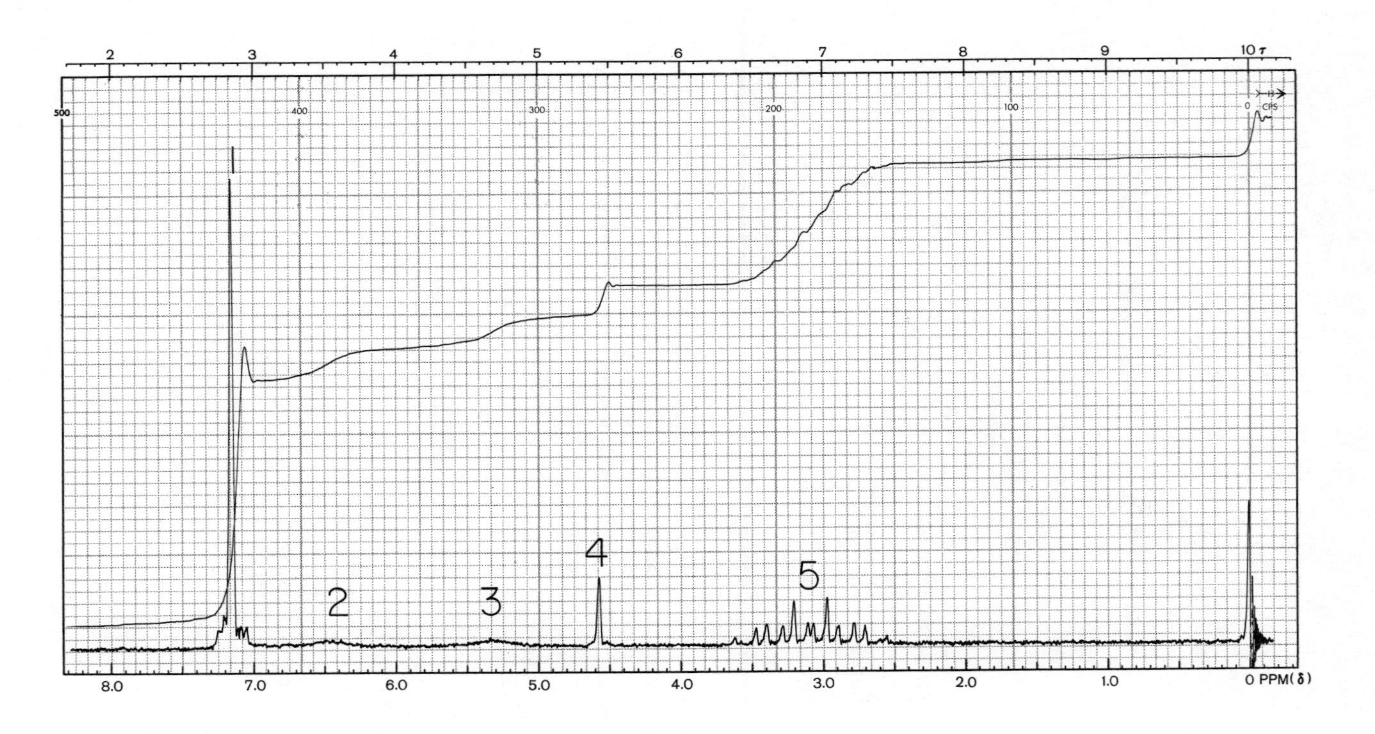
2
3
4
5
6
7
8
9
10 τ
500
400
300
200
100
0 CPS
1
2
3
4
5
8.0
7.0
6.0
5.0
4.0
3.0
2.0
1.0
0 PPM (δ)

$C_{16}H_{15}NO$

| Band | ① | ② | ③ | ④ | ⑤ |
|---|---|---|---|---|---|
| Integration | 100 | 14 | 13 | 13 | 50 |

---

*Questions*

1. Assign all of the numbered bands.
2. Explain why band ⑤ is so complex.
3. Why do the two $-NH_2$ protons have different chemical shifts?

Answers: p. 254.

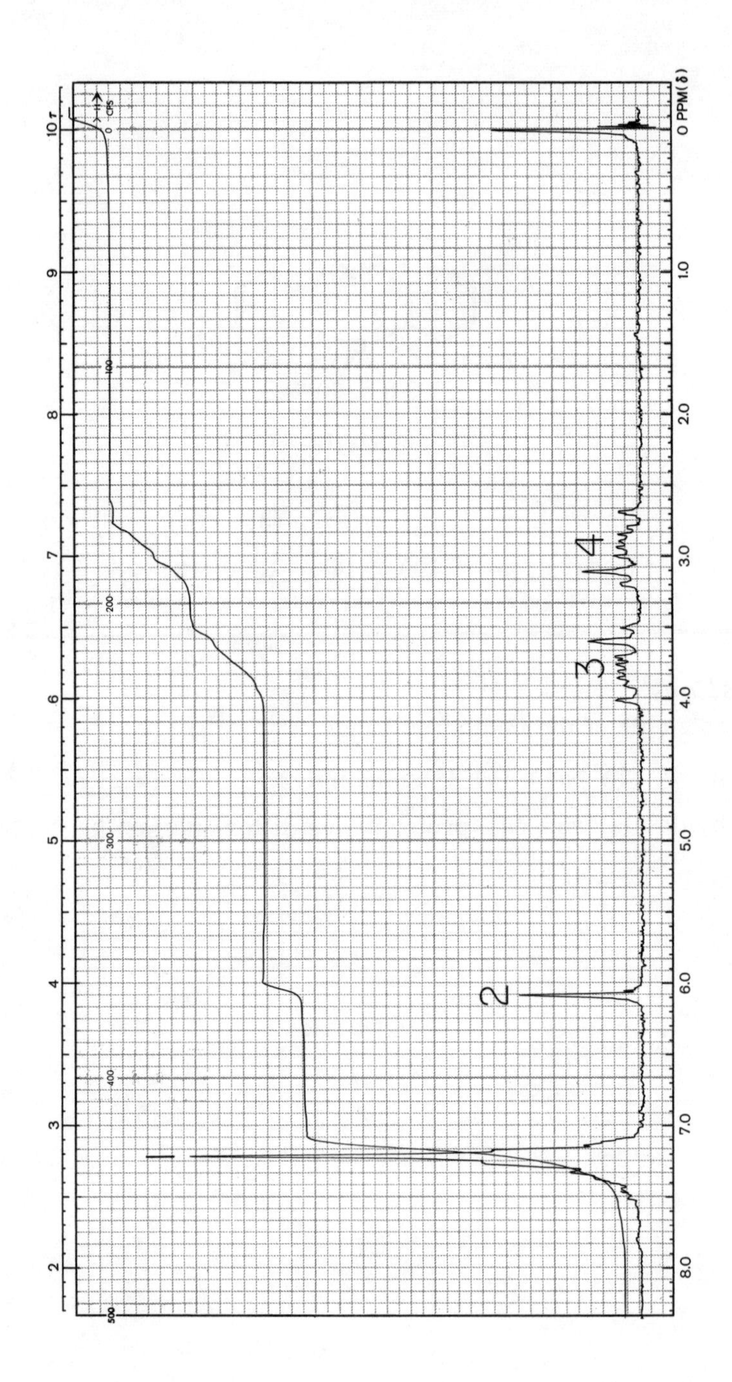

1
2
3
4

$C_{15}H_{13}Cl$

| Band | ① | ② | ③ | ④ |
|---|---|---|---|---|
| Integration | 134 | 16 | 30 | 33 |

None of the signals are removed by $D_2O$ exchange.

---

*Questions*

1. Propose a structure for this compound which is consistent with the spectrum.
2. Discuss the multiplicity of peaks in bands ③ and ④.

Answers: p. 255.

# SECTION 9

General problems.

*Suggested Reading*

Bhacca and Williams: pp. 63-123, 151-176, and 183-191.
Bible: pp. 113-117.
Emsley, Feeney, and Sutcliffe: pp. 442-476, 528-530, 740-749, 826-838, and 841-857.
Jackman: pp. 105-110.
Pople, Schneider, and Bernstein: pp. 98-99, 160-162, 233-296, 387-399, 402-417, and 422-441.

S. Sternhell reviews long-range coupling in Rev. Pure Appl. Chem. 14:15 (1964).

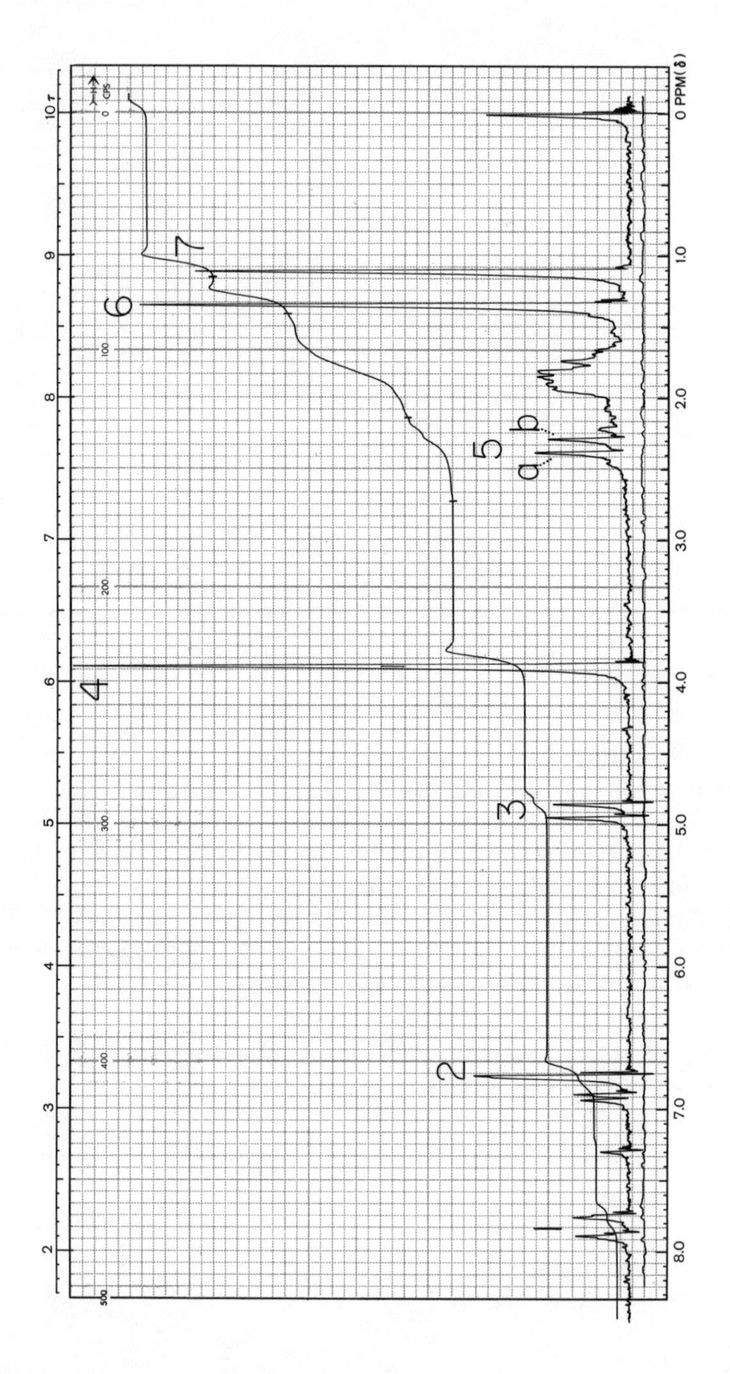

10 τ
9
8
7
6
5
4
3
2
0 CPS
100
200
300
400
500
0 PPM(δ)
1.0
2.0
3.0
4.0
5.0
6.0
7.0
8.0
7
6
5
a
b
4
3
2
1

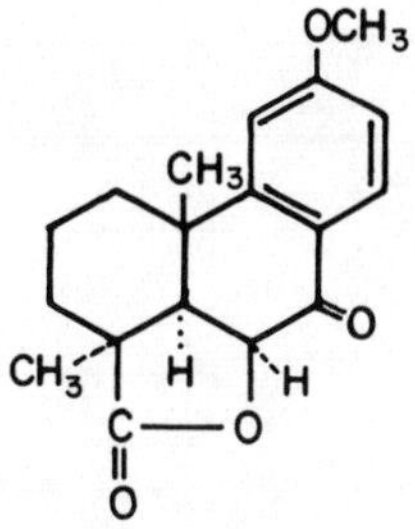

$C_{18}H_{20}O_4$

The lowest trace is the absorption curve between 1000 and 500 cps downfield from TMS.

| Band | ① | ② | ③ | ④ | ⑤ | ⑥ | ⑦ |
|---|---|---|---|---|---|---|---|
| Integration | 9 | 19.5 | 9 | 30 | 19 | 32 | 28 |

---

*Questions*

1. Assign the numbered bands to specific protons.
2. Describe the spin systems which give rise to the numbered bands.
3. What factors must be considered in predicting coupling in the AX pattern?

Answers: p. 256.

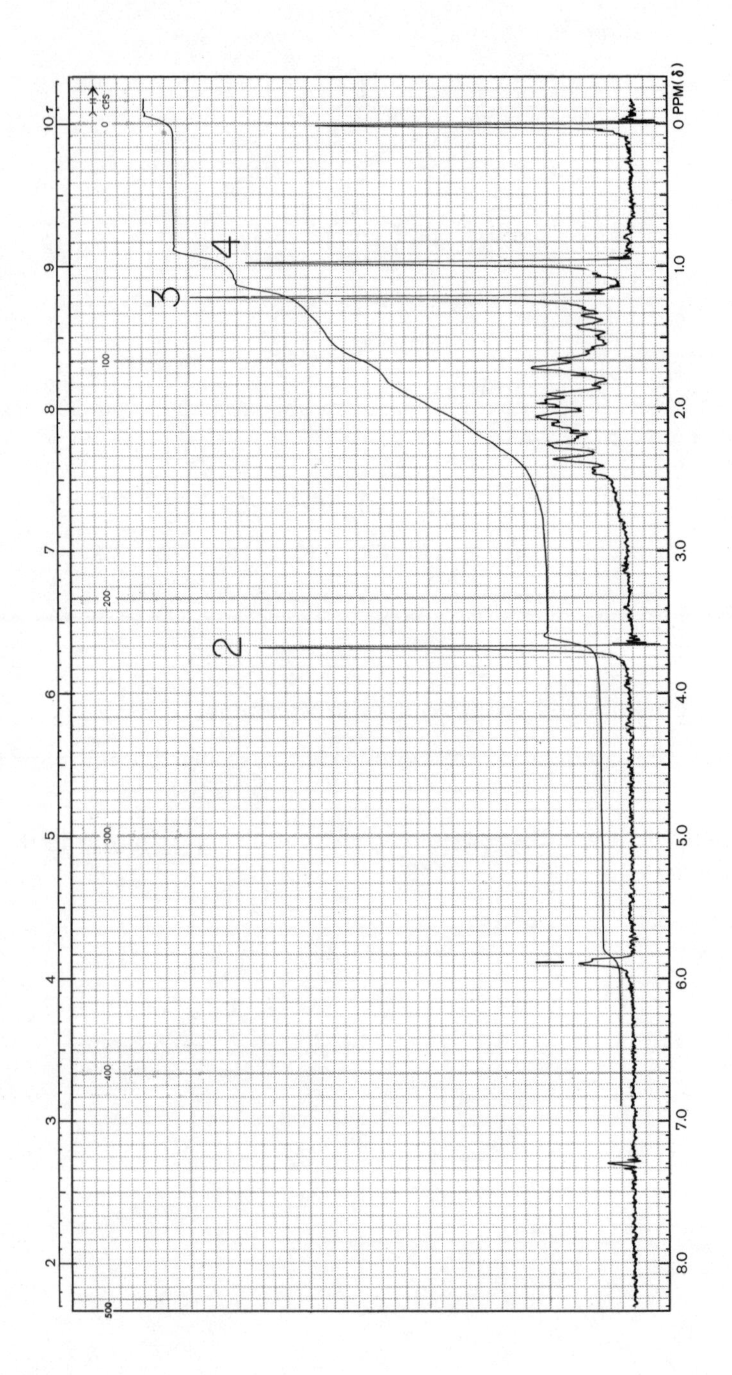

10 τ
9
8
7
6
5
4
3
2
0 CPS
100
200
300
400
500
0 PPM(δ)
1.0
2.0
3.0
4.0
5.0
6.0
7.0
8.0
1
2
3
4

This spectrum and the spectrum on p. 170 were determined using the following isomeric compounds:

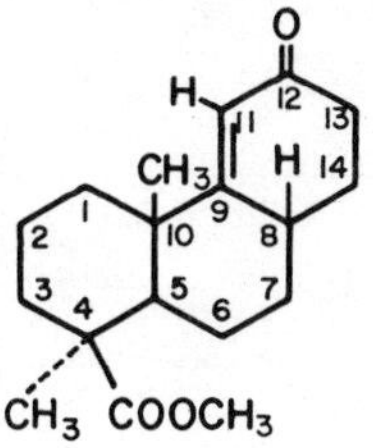

I 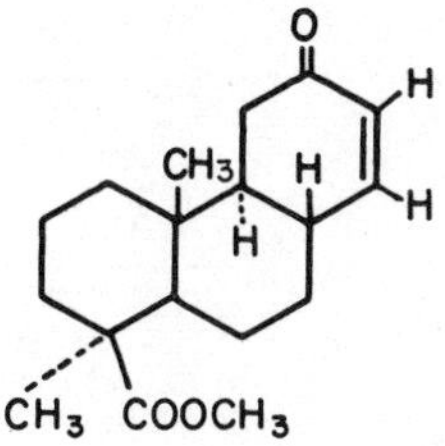 II

$C_{18}H_{26}O_3$

| Band | ① | ② | Total for spectrum |
| --- | --- | --- | --- |
| Integration | 6.5 | 21 | 185.5 |

On expansion, band ① in the spectrum at the left appears as a doublet separated by 1.5 cps. In the expanded curve, the peak at lower field is slightly less intense than the peak at higher field.

---

*Questions*

1. Identify each spectrum with its corresponding structure.
2. Assign the numbered bands in the spectrum above.
3. Was the instrument operating under optimum conditions for the recording of the spectrum shown on the left?

Answers: p. 256.

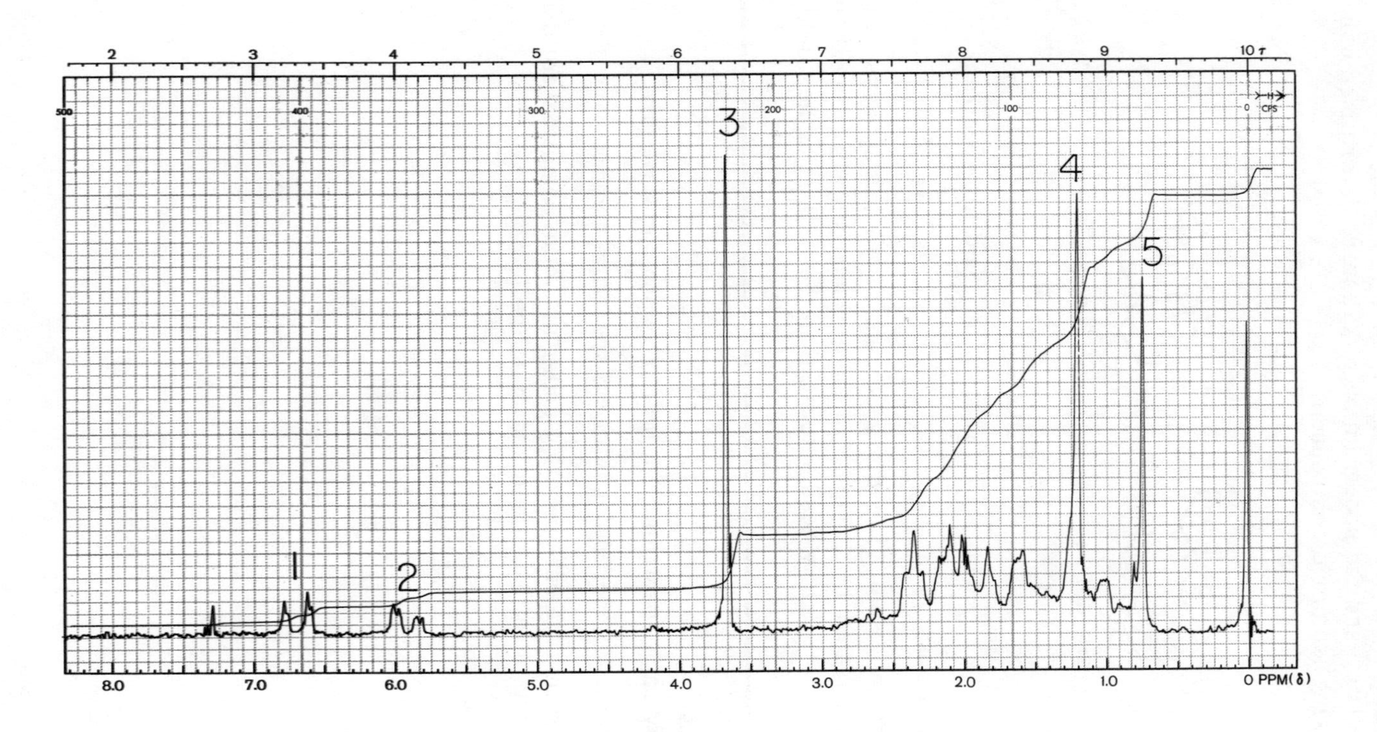

2
3
4
5
6
7
8
9
10 τ
500
400
300
200
100
0 CPS
1
2
3
4
5
8.0
7.0
6.0
5.0
4.0
3.0
2.0
1.0
0 PPM(δ)

This spectrum corresponds to one of the two structures given on p. 169. An expansion of the olefinic proton region is given on p. 172.

| Band | ① | ② | ③ | Total for spectrum |
| --- | --- | --- | --- | --- |
| Integration | 6.0 | 5.5 | 22.0 | 178 |

---

*Questions*

1. Assign the numbered bands.
2. What is unusual about the appearance of the multiplets in bands ① and ②?
3. By comparison with the expanded region shown on p. 172, propose a reason for the unusual appearance of the olefinic proton signals in the spectrum on the left.

Answers: p. 257.

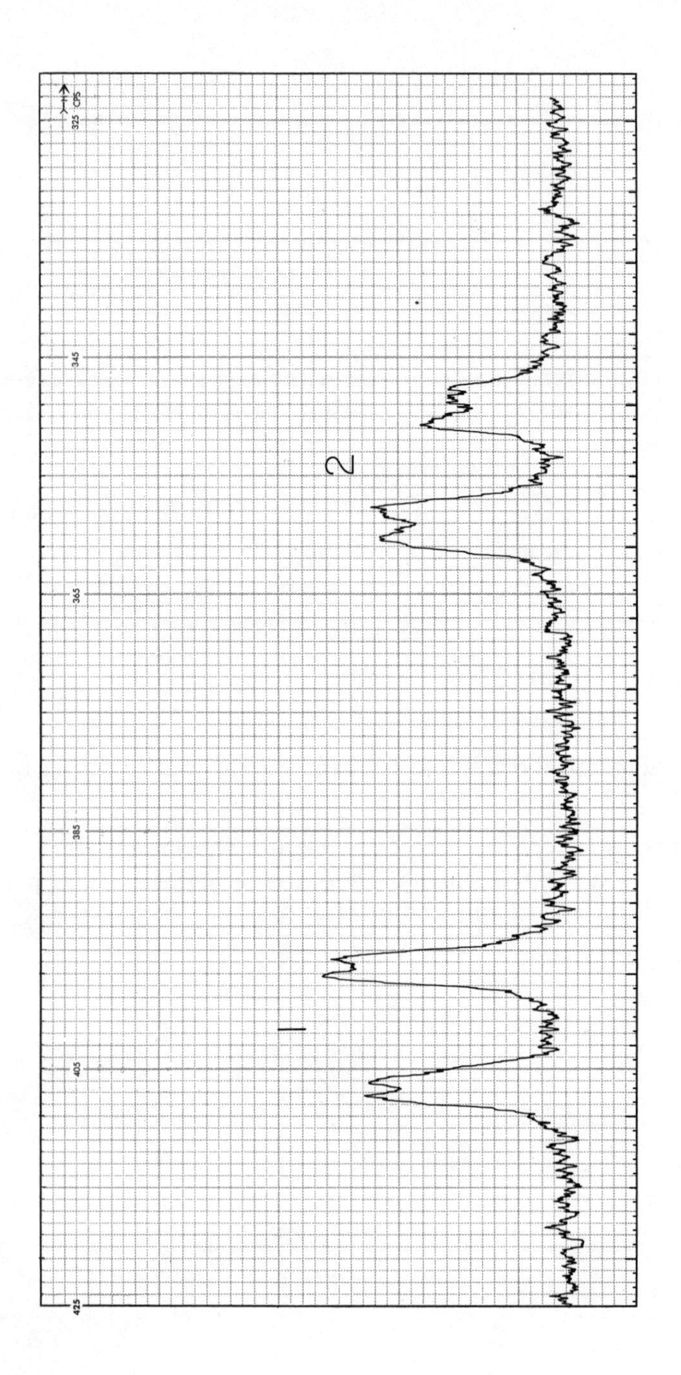
325 CPS
345
365
385
405
425
1
2

This is an expansion of the olefinic region of the spectrum on p. 170.

---

*Questions*

1. Propose couplings that would give this pattern.
2. Suggest a reason for the broadening of the peaks in band ②.
3. Upon what major factor does the magnitude of allylic proton–proton (CH=C—CH) coupling depend?

Answers: p. 258.

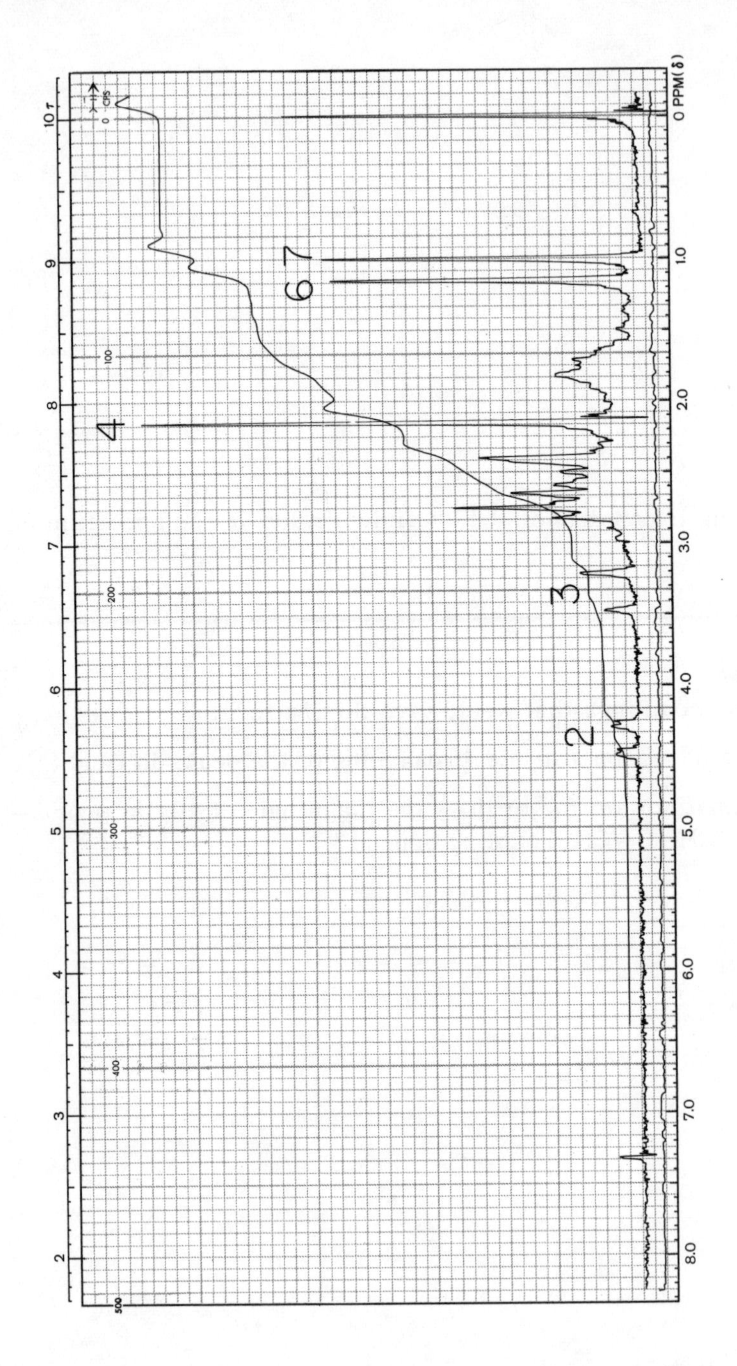
10 τ
9
8
7
6
5
4
3
2
100
200
300
400
500
CPS
0 PPM (δ)
1.0
2.0
3.0
4.0
5.0
6.0
7.0
8.0
4
6 7
3
2

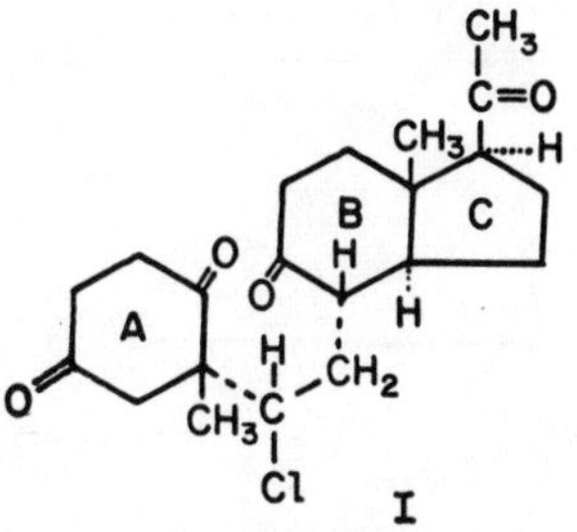

CH3
C=O
CH3 ···H
B C
H
O O H
A
O H CH2
CH3C
Cl I

$C_{21}H_{29}ClO_4$

The spectrum of the very closely related compound (II) is shown on p. 176. Corresponding bands are numbered identically on each of the two spectra. The lowest trace on each spectrum is the scan from 1000 to 500 cps.

---

*Questions*

1. Assign as many of the peaks as possible using the spectra of these two related compounds.
2. Explain the presence of the doublet in band ③.
3. Is the separation of the peaks in band ③ consistent with the explanation given?
4. What is the approximate distance from the low-field peak in band ③ to the corresponding high-field peak in the other part of the pattern?

Answers: p. 259.

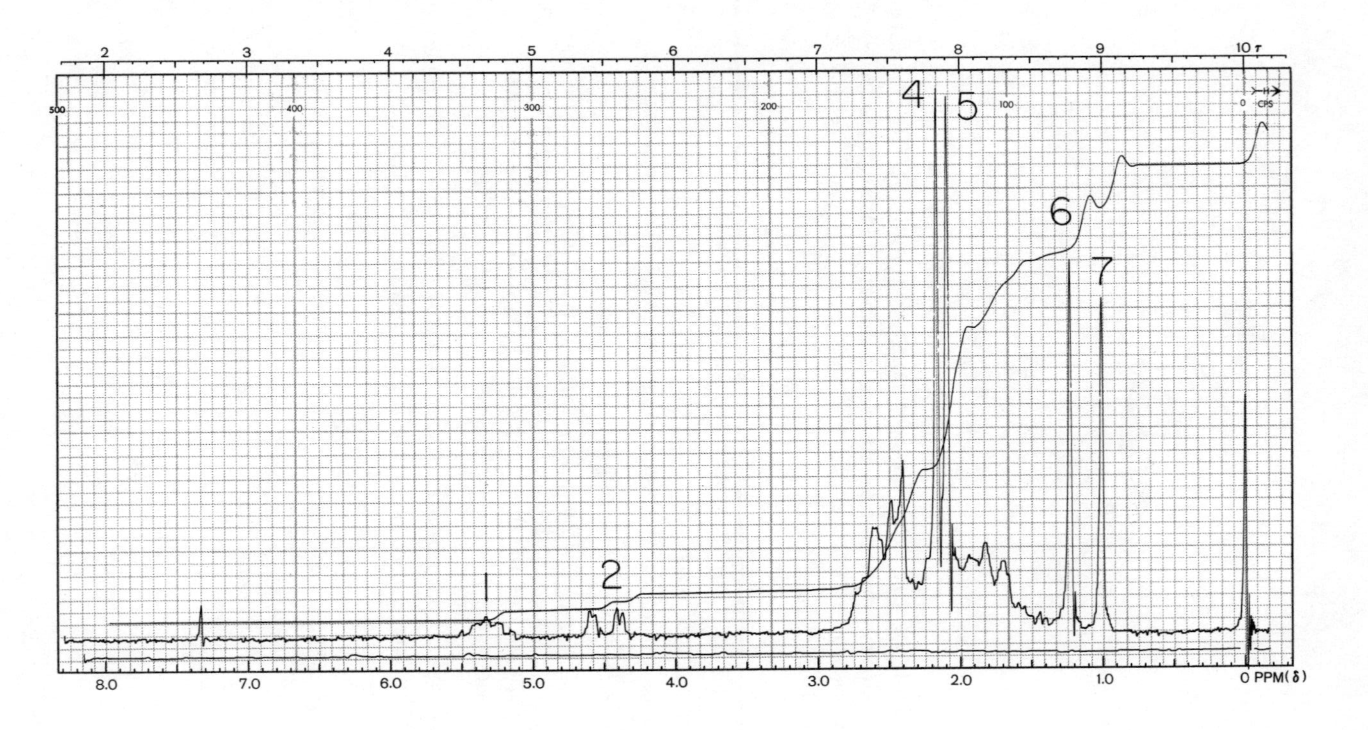

10 τ
9
8
7
6
5
4
3
2
0 PPM(δ)
1.0
2.0
3.0
4.0
5.0
6.0
7.0
8.0
100
200
300
400
500
1
2
4
5
6
7

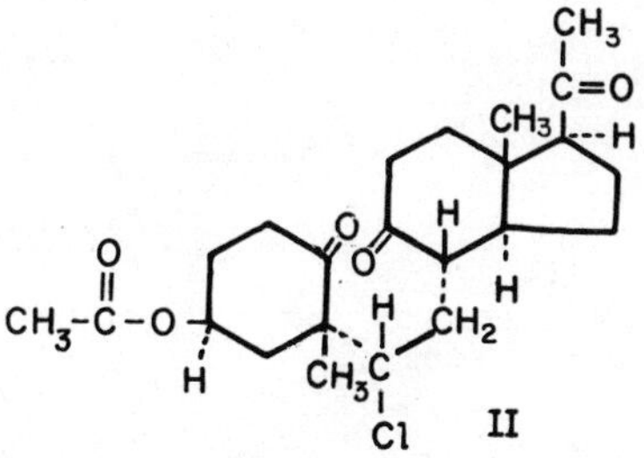

II

$C_{23}H_{33}ClO_5$

The spectrum of a closely related derivative is shown on p. 174.

---

*Questions*

1. Can the separations (11.5 and 2.5 cps) in band ② be assumed to be coupling constants?

2. What tests might help to determine if the smaller splittings in band ② represent a true coupling constant or a splitting due to "virtual" coupling?

3. What can be said about the coupling between the proton which causes band ② and neighboring protons?

Answers: p. 260.

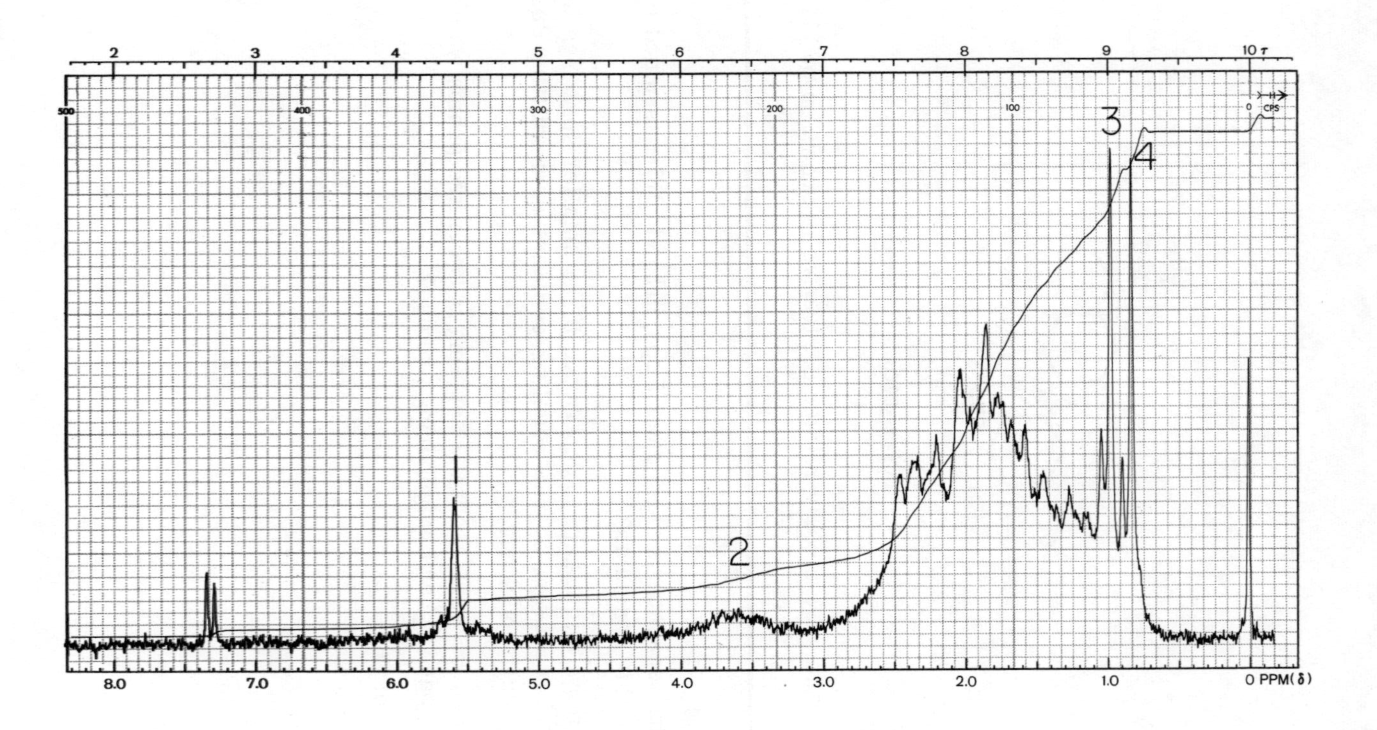
2
3
4
5
6
7
8
9
10 τ
500
400
300
200
100
0 CPS
3
4
1
2
8.0
7.0
6.0
5.0
4.0
3.0
2.0
1.0
0 PPM (δ)

This is the spectrum, taken after $D_2O$ exchange, of the product which resulted from an attempted preparation of the following compound:

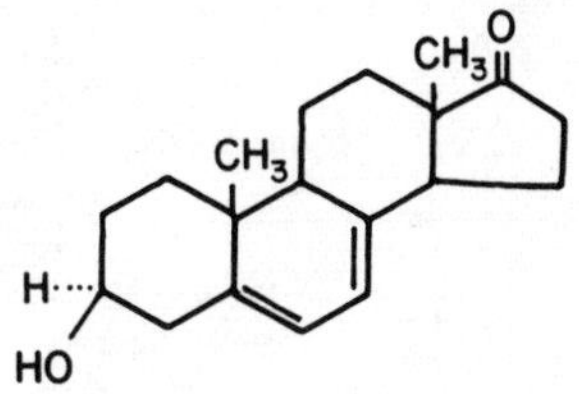

$C_{19}H_{26}O_2$

---

*Questions*

1. Is the sample pure?
2. Indicate one solvent used in the reaction or in the work-up.
3. Assign the numbered bands.
4. Explain the appearance of band ①.

Answers: p. 260.

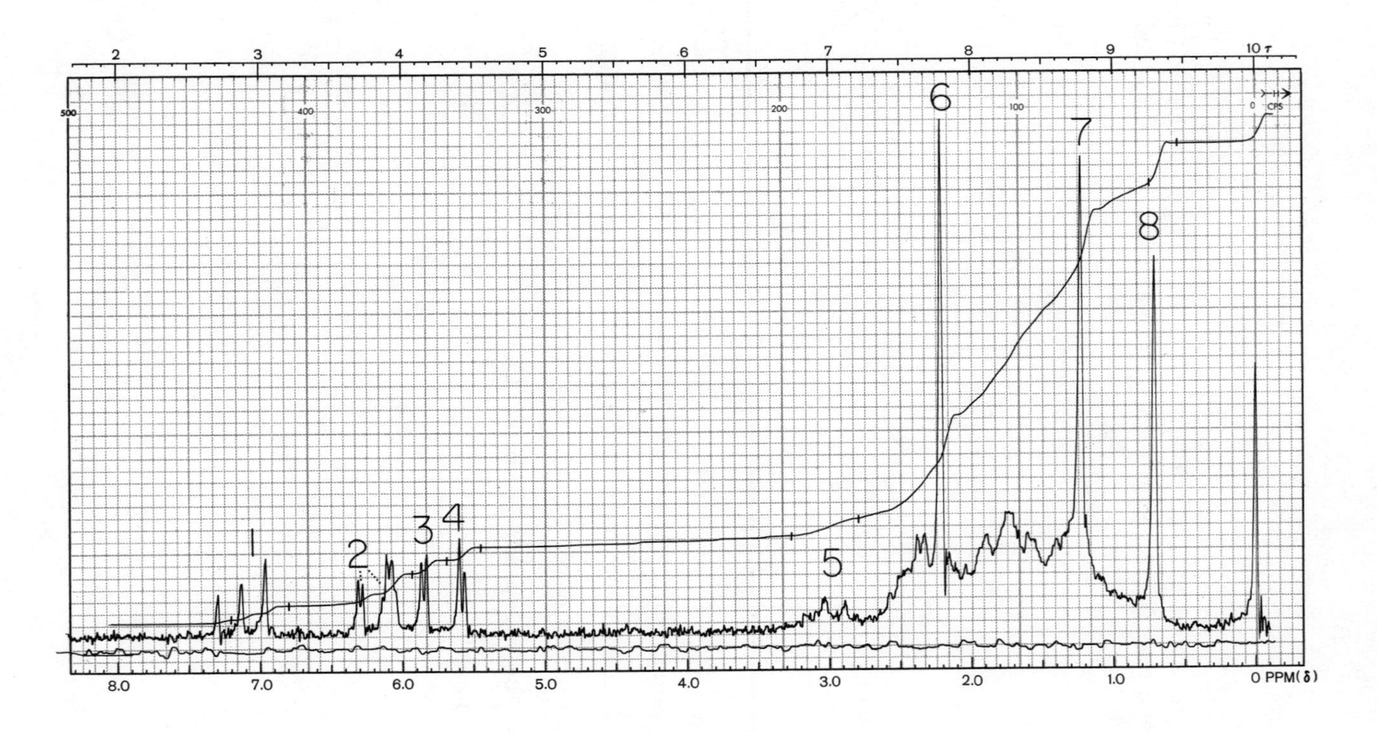
2
3
4
5
6
7
8
9
10 τ
500
400
300
200
100
0 CPS
1
2
3
4
5
6
7
8
8.0
7.0
6.0
5.0
4.0
3.0
2.0
1.0
0 PPM(δ)

This is the spectrum of one of the following isomers:

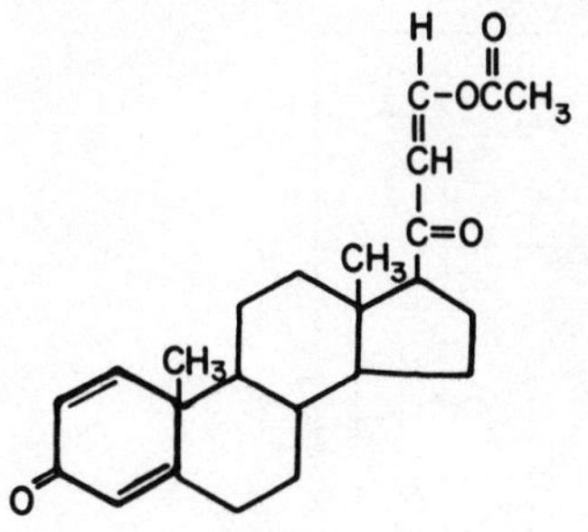

I

$C_{24}H_{30}O_4$

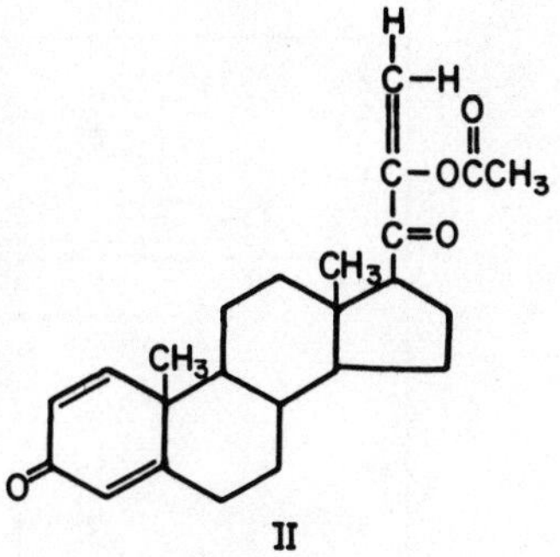

II

The lower trace is the absorption curve over the region 1000 – 500 cps.

| Band | ① | ② | ③ | ④ | ⑤ | ⑧ | Total for spectrum |
|---|---|---|---|---|---|---|---|
| Integration | 5.5 | 12.5 | 5.5 | 5.5 | 7 | 17 | 200 |

---

*Questions*

1. Assign the numbered bands to specific protons.
2. Which of the two structures is correct?

Answers: p. 261.

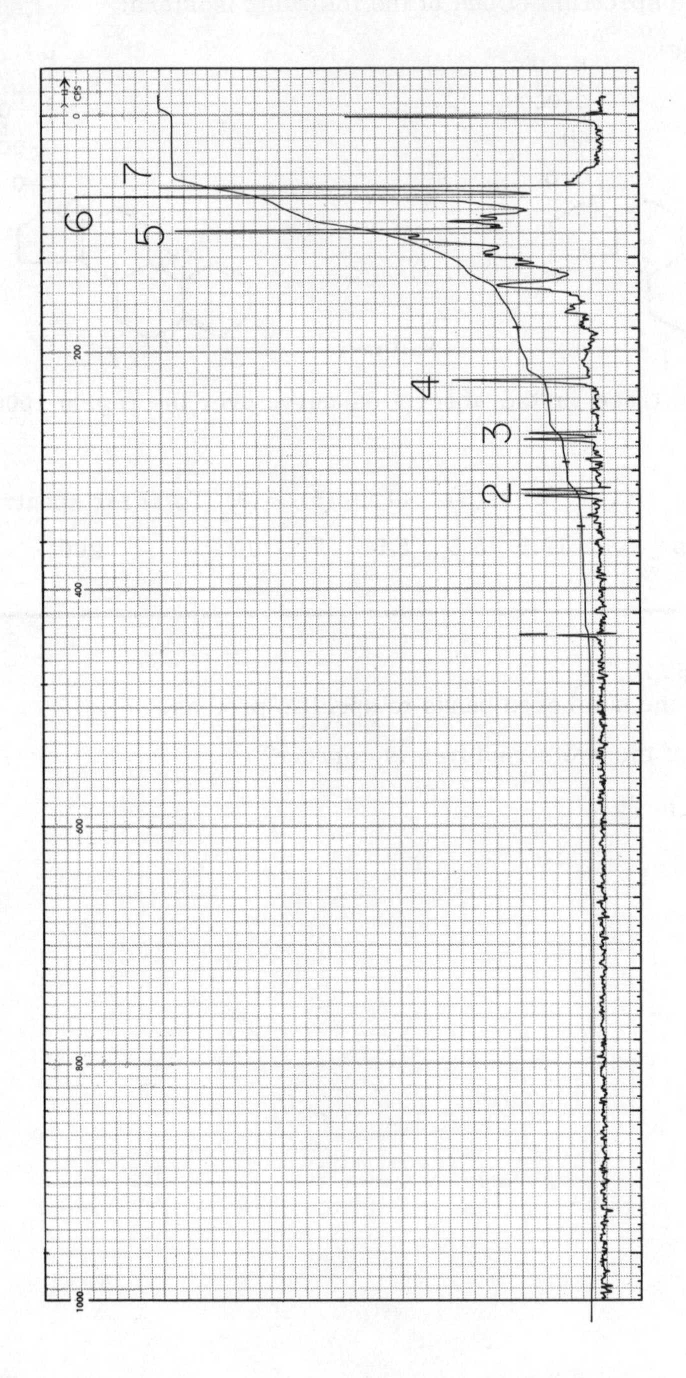

0 CPS
200
400
600
800
1000
7
6
5
4
3
2

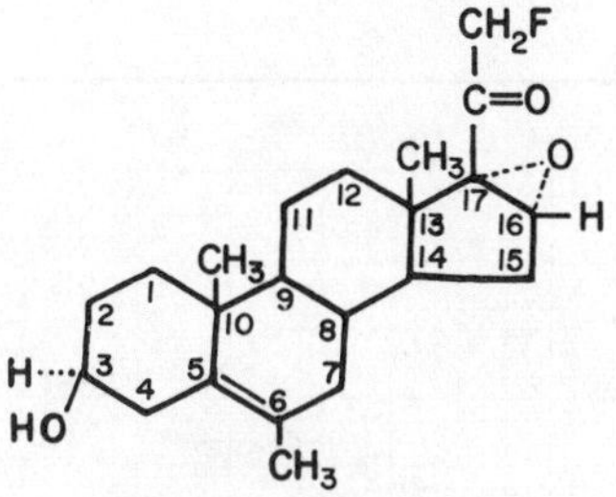

$C_{22}H_{31}FO_3$

This spectrum shows the region from 1000 to 0 cps. None of the numbered peaks are removed by $D_2O$ exchange. An amplified expansion of bands ②, ③, and ④ is shown on p. 184.

| Band | ① | ② | ③ | ④ | Total for spectrum |
|---|---|---|---|---|---|
| Integration | 1 | 7 | 7 | 13.5 | 175.5 |

---

*Questions*

1. Assign the numbered peaks to specific protons.
2. Explain the splittings of bands ② and ③ and the sharpness of the peak in band ④.

Answers: p. 262.

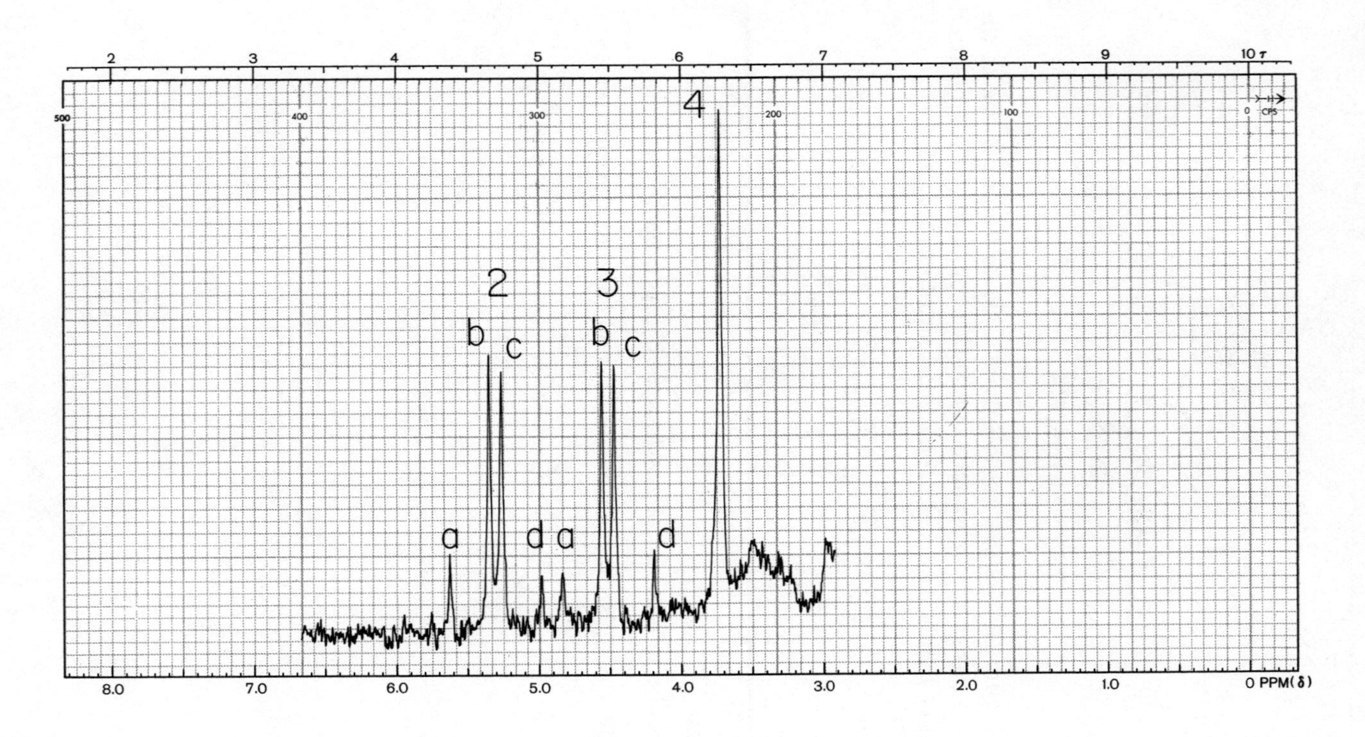

2
3
4
5
6
7
8
9
10 τ
500
400
300
200
100
0 CPS
4
2
b c
3
b c
a
d a
d
8.0
7.0
6.0
5.0
4.0
3.0
2.0
1.0
0 PPM(δ)

This is the amplified expansion of bands ②, ③, and ④ in the spectrum on p. 182. The total region covered by the chart (not the trace) is 500 to 0 cps.

| Peak | Position (cps) |
|---|---|
| 2a | 337.5 |
| b | 321 |
| c | 315.5 |
| d | 299 |
| 3a | 290 |
| b | 273 |
| c | 268 |
| d | 351 |

---

*Questions*

1. What information can be determined from bands ② and ③?
2. What would be the appearance of the fluorine spectrum?

Answers: p. 263.

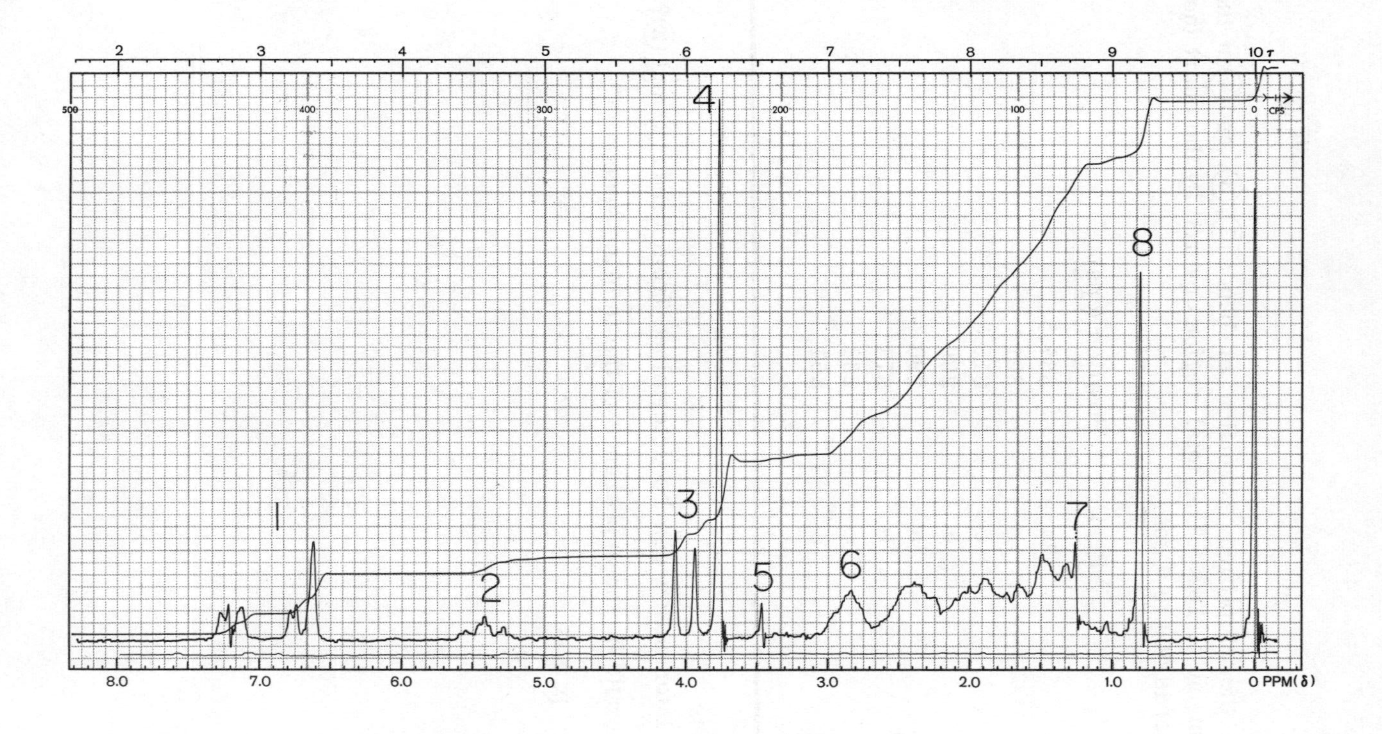

2
3
4
5
6
7
8
9
10 τ
500
400
300
200
100
0
CPS
1
2
3
4
5
6
7
8
8.0
7.0
6.0
5.0
4.0
3.0
2.0
1.0
0 PPM(δ)

This is the spectrum of the crude product obtained from the following reaction:

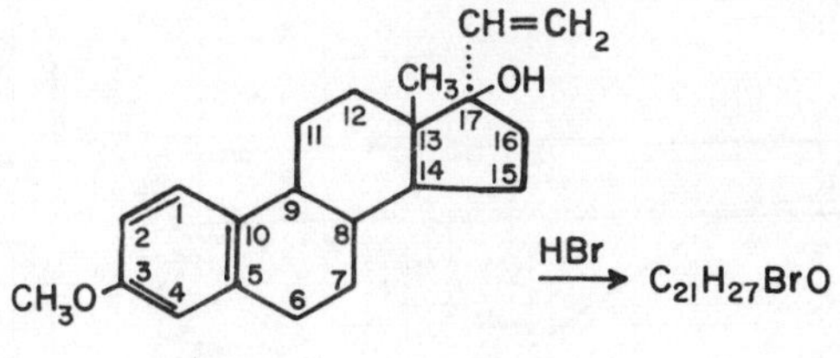


The lower tracing shows the region 1000 to 500 cps. Only peak ⑤ is removed by $D_2O$ exchange.

| Band | ① | ② | ③ | ④ | ⑤ | ⑥ | ⑧ | Total for spectrum |
|---|---|---|---|---|---|---|---|---|
| Integration | 25 | 6 | 15 | 24 | 1.5 | 17 | 23.5 | 224 |

*Questions*

1. What solvent was probably used in the work-up? What simple test would help to confirm this?
2. Propose a structure for the product which is consistent with the spectrum.
3. Does this spectrum solve all of the stereochemical problems in the proposed structure?

Answers: p. 264.

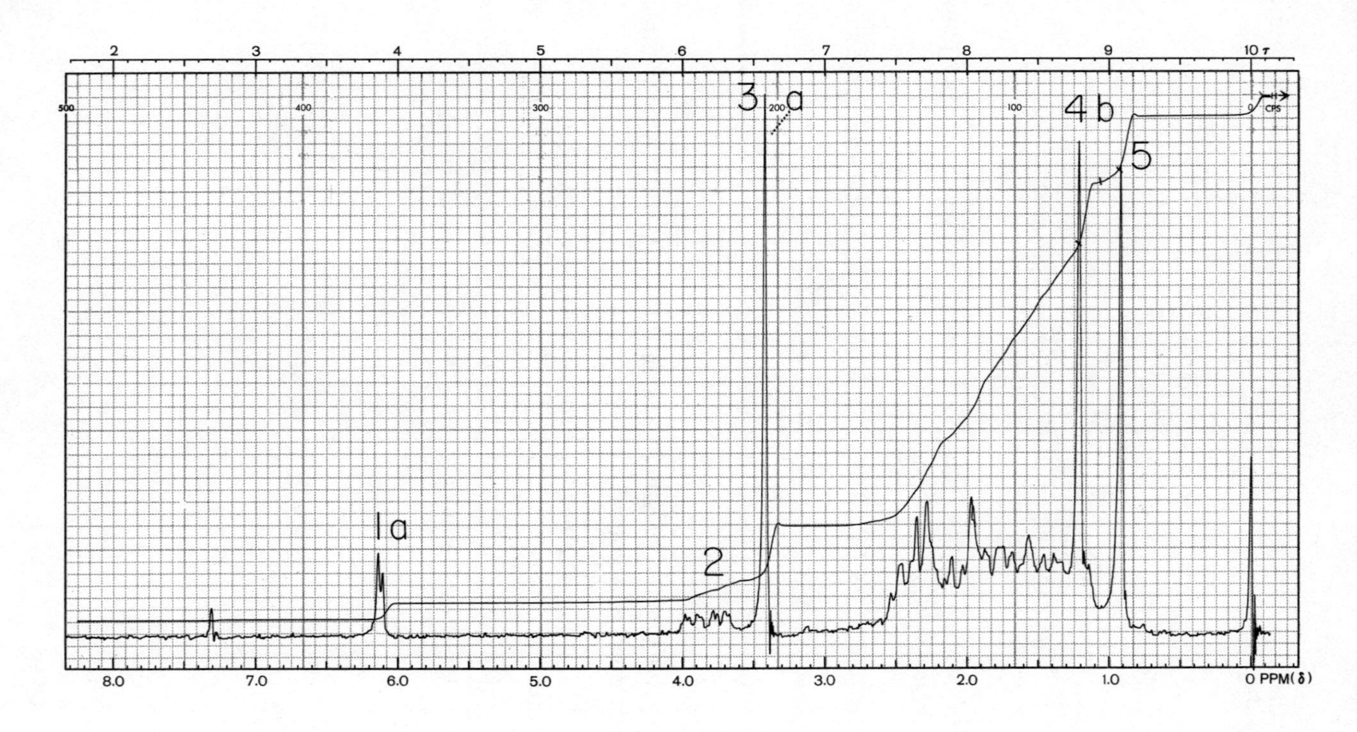

2
3
4
5
6
7
8
9
10 τ
500
400
300
200
100
0
CPS
1a
2
3
a
4
b
5
8.0
7.0
6.0
5.0
4.0
3.0
2.0
1.0
0 PPM(δ)

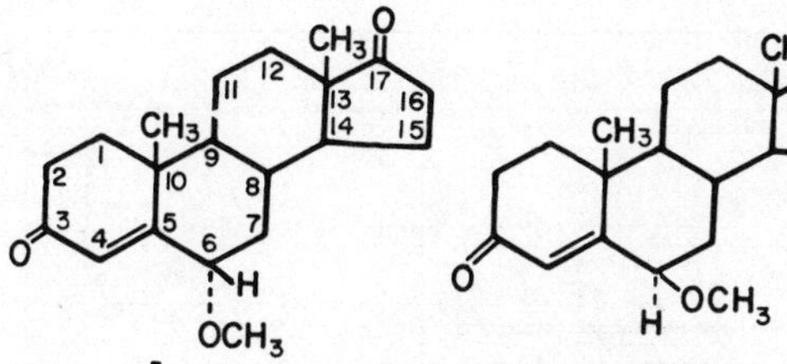

I

II

$C_{20}H_{28}O_3$

One of the two isomeric compounds was isolated in pure form. The spectrum of this pure isomer and one of a mixture of the two isomers were obtained.

| Band | ① | ② | ③ | ④b | ⑤ | Total for spectrum |
|---|---|---|---|---|---|---|
| Integration | 6.5 | 8.5 | 23 | 25 | 23 | 212 |

---

*Questions*

1. Does the spectrum at the left or the one on p. 190 belong to the pure isomer?
2. Which of the two isomers was obtained pure?

Answers: p. 265.

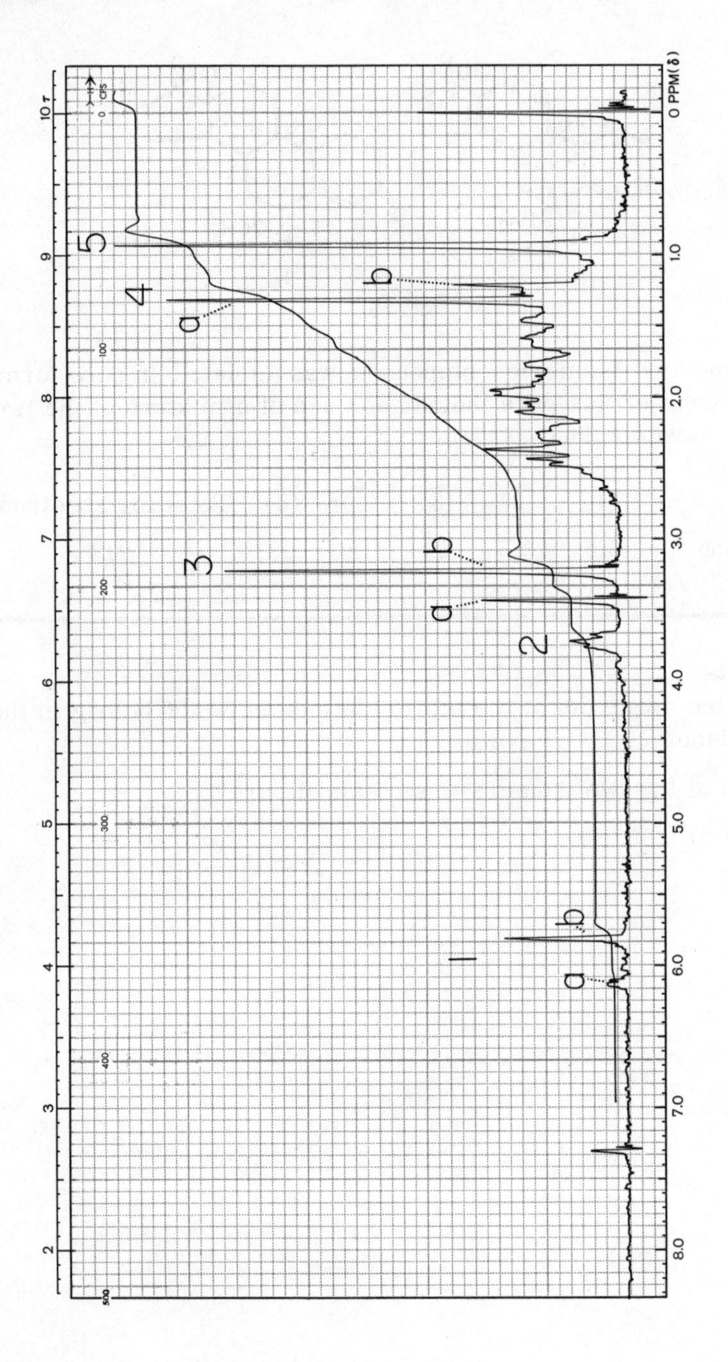

0 PPM(δ)
1.0
2.0
3.0
4.0
5.0
6.0
7.0
8.0
10 τ
9
8
7
6
5
4
3
2
CPS
100
200
300
400
500
5
4
3
2
1
a
b

This spectrum was introduced on p. 188.

| Band | ①a + b | ② | ③a | ③b |
|---|---|---|---|---|
| Integration | 7.5 | 8.5 | 7 | 14.5 |

---

*Question*

1. What is the approximate molar composition of this mixture?

Answer: p. 266.

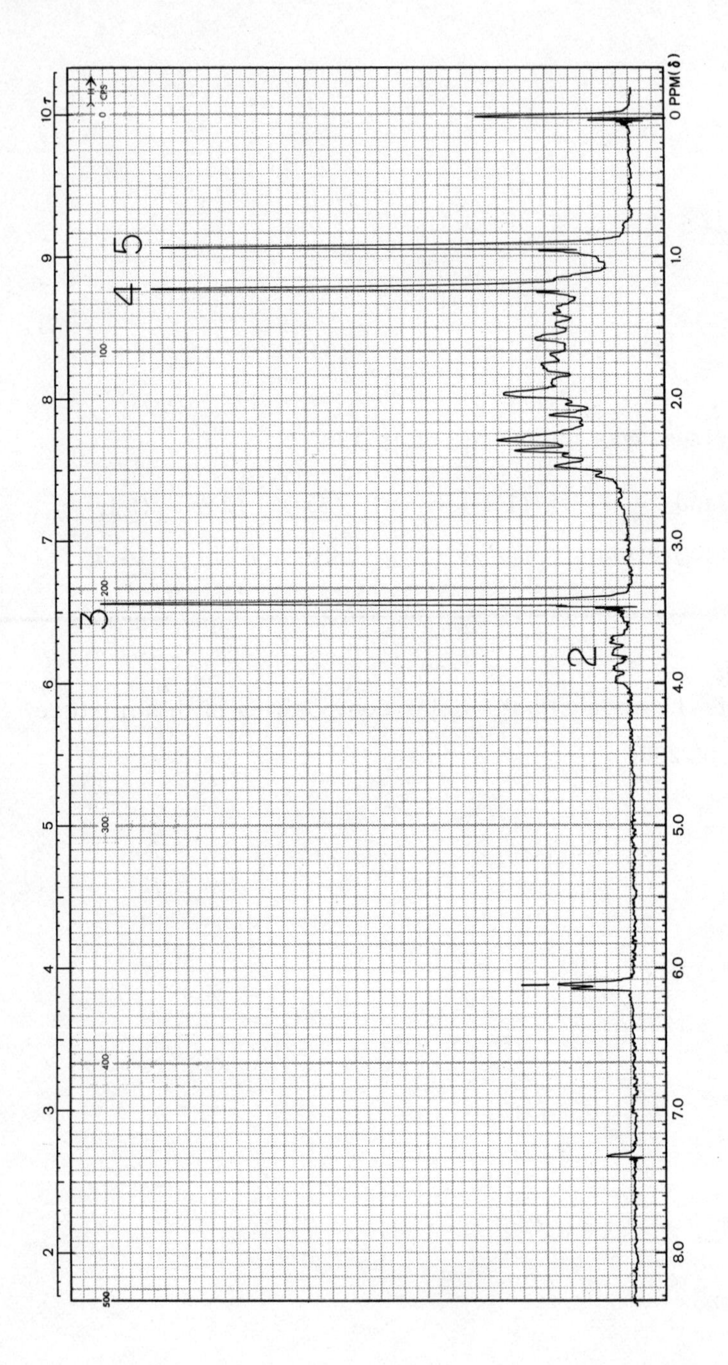

10 τ
9
8
7
6
5
4
3
2
0 PPM(δ)
1.0
2.0
3.0
4.0
5.0
6.0
7.0
8.0
0
CPS
100
200
300
400
500
2
3
4
5

This spectrum was obtained using the same solution used for the spectrum on p. 188. The spectrum on the left was swept from high field to low field, however.

---

*Question*

1. What two differences were caused by the reverse sweeping?

Answer: p. 267.

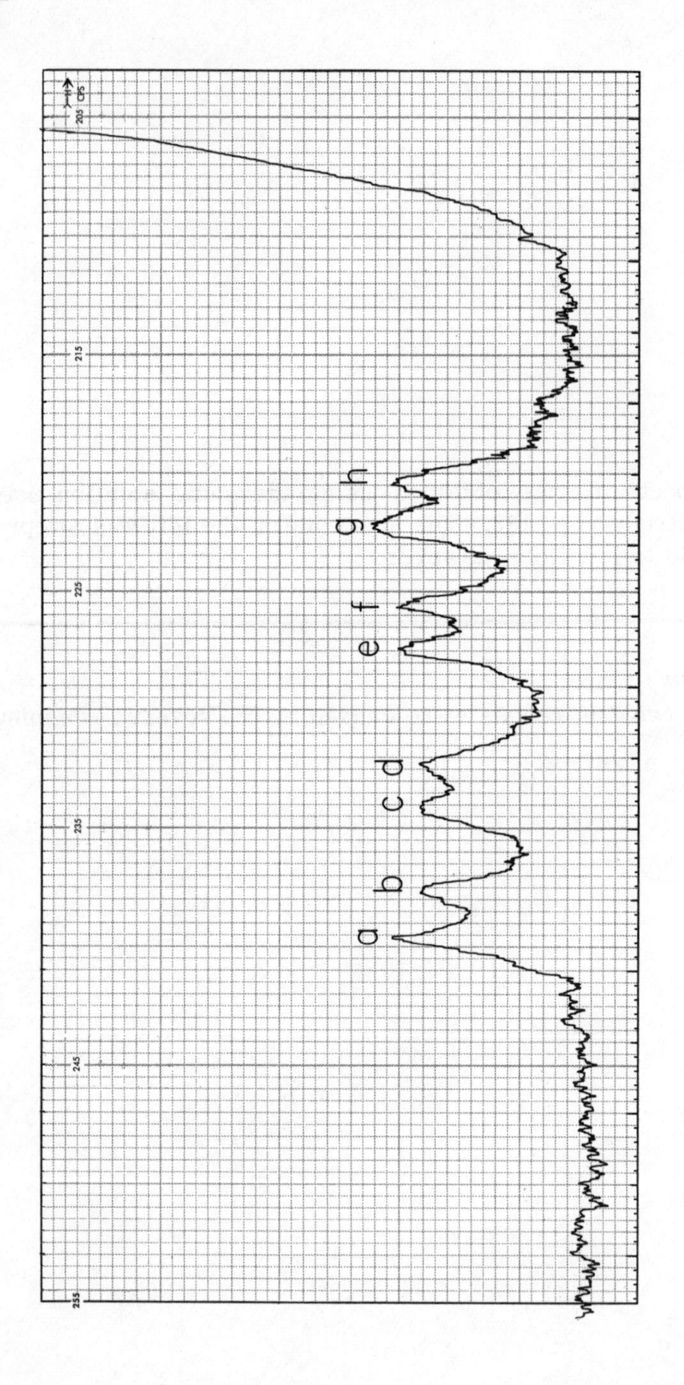

205 CPS
215
225
235
245
255
a
b
c
d
e
f
g
h

This is an expansion of band (2) in the spectrum on p. 188. The total width of the spectrum is 50 cps. Part of the tall signal due to the methoxy protons is seen on the high-field side of the spectrum.

---

*Questions*

1. Describe the spin system involved in this pattern.
2. Which of the splittings can be taken as coupling constants?

Answers: p. 267.

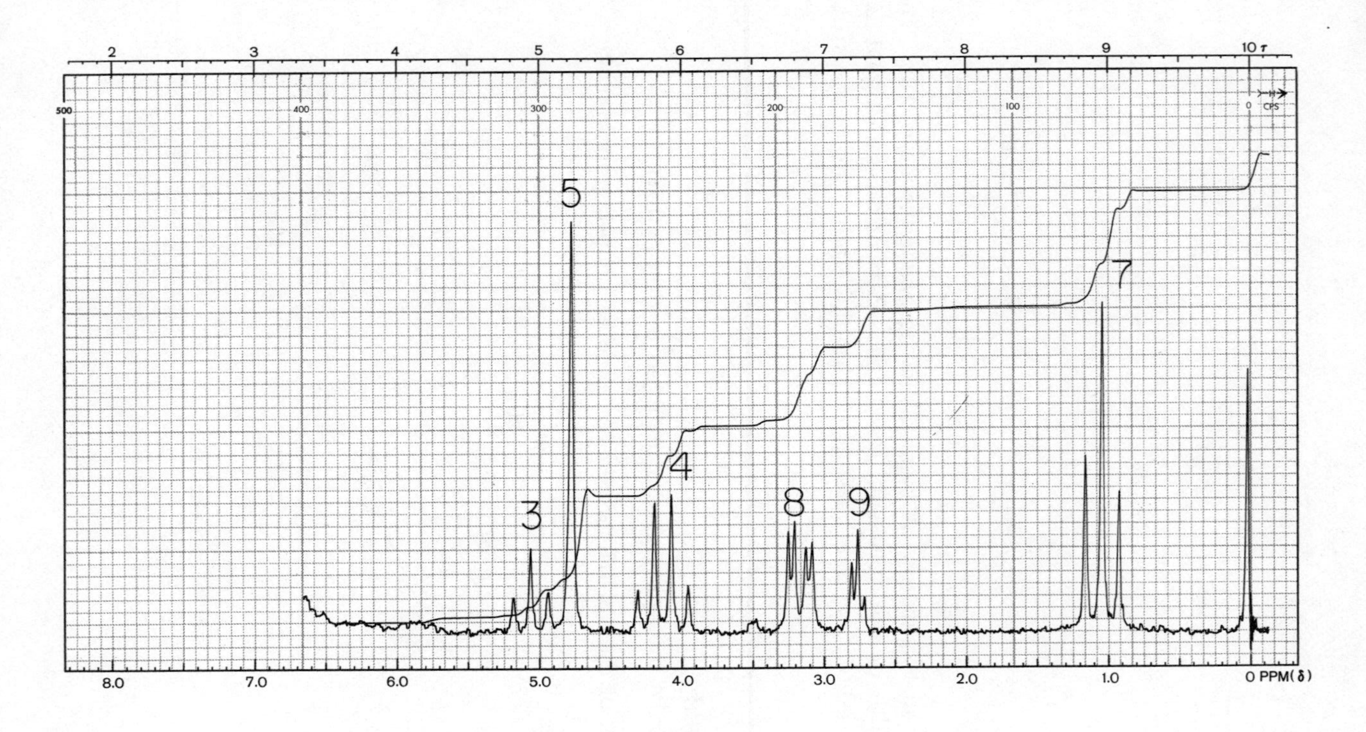

2
3
4
5
6
7
8
9
10 τ
500
400
300
200
100
0 CPS
5
7
4
3
8
9
8.0
7.0
6.0
5.0
4.0
3.0
2.0
1.0
0 PPM(δ)

$$C_6H_5-\overset{\overset{O}{\|}}{C}-\underset{\underset{CH_2C\equiv CH}{|}}{\overset{\overset{H}{|}}{C}}-\overset{\overset{O}{\|}}{C}OCH_2CH_3$$

$C_{14}H_{14}O_3$

This spectrum was determined in pyridine. The region between 500 and 400 cps (8.33 – 6.67 ppm) was omitted because of the absorption due to the protons in the solvent. The spectrum on p. 198 was obtained using the same compound dissolved in deuterodimethylsulfoxide ($CD_3\overset{\overset{O}{\|}}{S}CD_3$). Corresponding bands in the two spectra have been given the same numbers.

| Band | ③ | ④ | ⑤ | ⑦ | ⑧ | ⑨ |
|---|---|---|---|---|---|---|
| Integration | 15 | 29 | 33 | 48 | 31 | 15.5 |

---

*Questions*

1. Assign the bands to specific protons.
2. Determine as many coupling constants as possible from this spectrum.

Answers: p. 268.

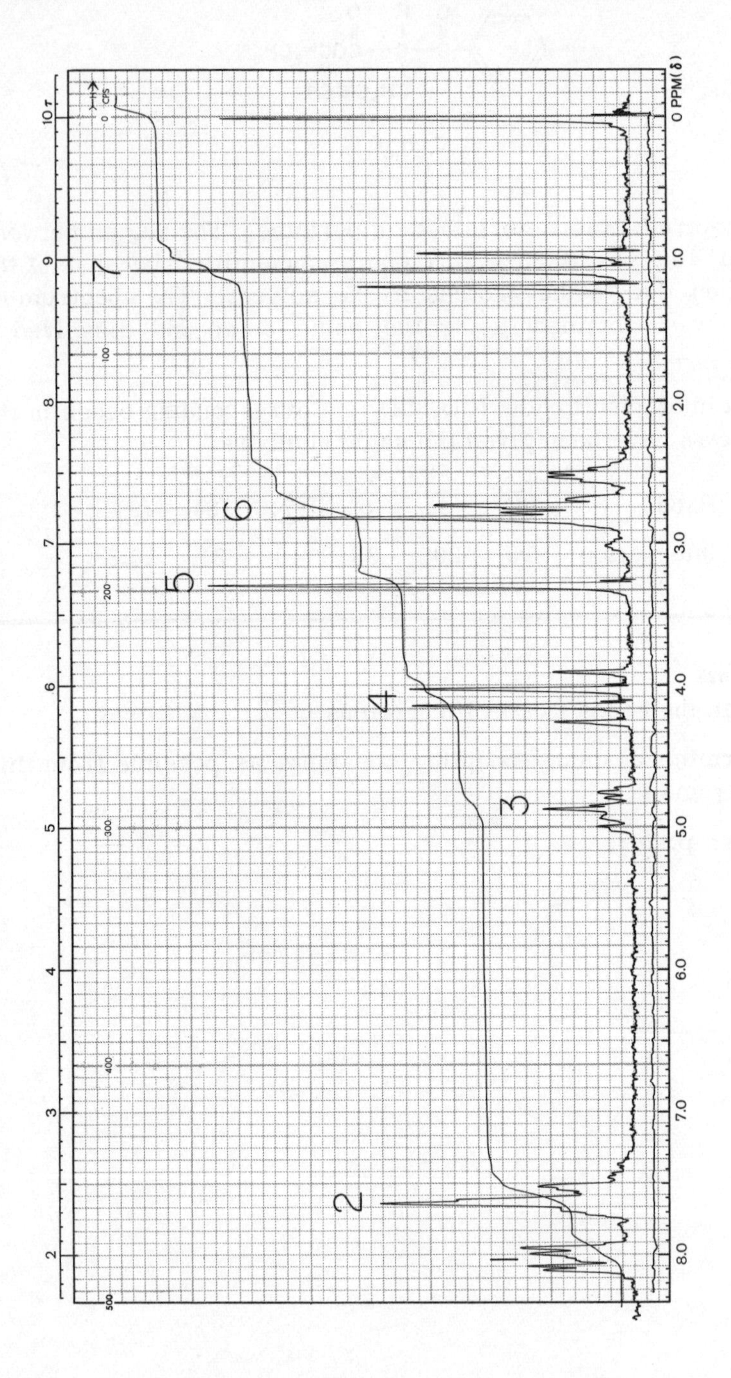
1
2
3
4
5
6
7
10 τ
0 PPM (δ)
CPS

$$C_6H_5\text{-}\overset{O}{\overset{\|}{C}}\text{-}\underset{CH_2C\equiv CH}{\overset{H}{C}}\text{-}\overset{O}{\overset{\|}{C}}\text{-}OCH_2CH_3$$

$C_{14}H_{14}O_3$

This spectrum was determined in deuterodimethylsulfoxide ($CD_3\overset{O}{\overset{\|}{S}}CD_3$). The lower trace covers the region 1000 to 500 cps. The spectrum on p. 196 was obtained using the same compound dissolved in pyridine.

| Band | ① | ② | ③ | ④ | ⑤ | ⑥ | ⑦ |
|---|---|---|---|---|---|---|---|
| Integration | 21 | 35 | 10 | 23 | 17 | 35 | 35.5 |

---

*Questions*

1. Assign the bands to specific protons.
2. Discuss the appearance of bands ③ and ⑥.
3. Suggest ways for testing the explanations given in the answer to question 2 above.

Answers: p. 268.

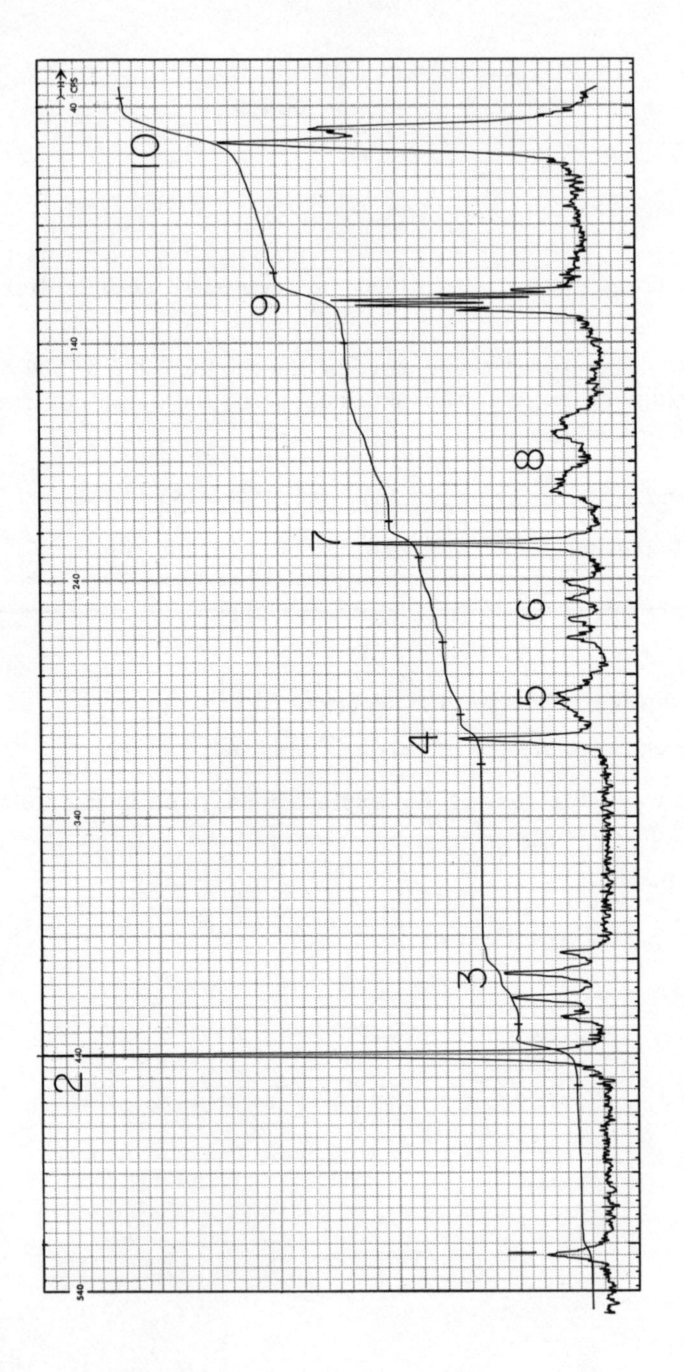

40
CPS
140
240
340
440
540
1
2
3
4
5
6
7
8
9
10

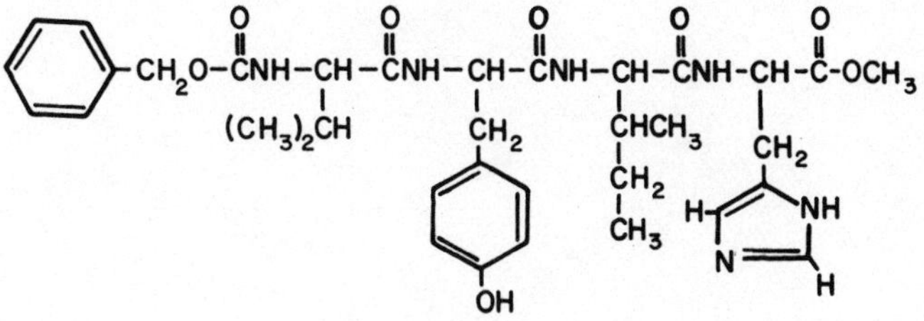

$C_{35}H_{46}N_6O_8$

This spectrum, which was determined in $CD_3COOD$, is offset 40 cps.

| Band | ① | ② | ③ | ④ | ⑤ | ⑥ | ⑦ | ⑧ | ⑨ | ⑩ | Total for spectrum |
|---|---|---|---|---|---|---|---|---|---|---|---|
| Integration | 4.2 | 24 | 15 | 9 | 8 | 9 | 12.5 | 18 | 29 | 64 | 196 |

---

*Questions*

1. It was not possible to adjust the instrument to obtain good ringing of the sharp signals due to the protons in the molecule. Suggest a reason for this difficulty.
2. Assign all of the numbered peaks to specific protons.

Answers: p. 271.

# ANSWERS

*Answers for Page* 7

1. Scan B was determined under the best operating conditions. Note that the peaks in scan A dip down below the base line on the left side and rise higher than the base line on the right side. This is due to poor "phasing." The peaks in scan C are broad and show no ringing. This was the result of a magnetic field which was not homogeneous over the entire sample.

2. The chloroform proton signal appears at lower magnetic field because of the electron-withdrawing property of the chlorine atoms. Electrons tend to circulate around a nucleus in the direction that opposes the applied magnetic field. The electron circulation shields the nucleus from the effect of the external magnetic field. This means that a higher magnetic field has to be applied in order to bring the nucleus into resonance. The attachment of electron-withdrawing groups "deshields" by decreasing the electron density around the proton, thus enabling the proton to "see" more of the applied magnetic field and causing the signal to appear at a lower applied field.

3. The separation of the two sharp peaks is called the difference in chemical shifts. This separation (435.5 cps at 60 Mcps) can be used to check the calibration of the instrument. Peak positions are always measured as separations from a reference peak. Here, the reference peak is the signal of the twelve protons in $(CH_3)_4Si$. Note that this signal has been adjusted to 0 cps in scan B. The reference compound used here is dissolved in the solution and is, thus, called an internal reference. Since it is understood that a reference signal is always employed, the position of a sharp peak is usually simply called the chemical shift.

4. The chemical shift can be expressed in delta ($\delta$) units by dividing the shift expressed in cps by the frequency used. Thus, the chloroform peak is at 435.5 cps or at 435.5 cps/60 Mcps or 7.26 ppm downfield from internal TMS. To obtain the answer in tau units, the position expressed in ppm is subtracted from 10, which is the position assigned to TMS. Thus, the chloroform peak would be at 10 – 7.26 or 2.74 ppm (expressed in tau units). All three scales appear on most of the spectra in this book.

5. The chemical shift expressed in tau or delta units is independent of the frequency used. Thus, at any frequency the chloroform peak would be at delta equal to 7.26 ppm or tau equal to 2.74 ppm. The chemical shift in cps is proportional to the frequency used, so that at 100 Mcps the chloroform peak would be at 435.5 (100/60) or at 726 cps. Note that this is equal to the position on the delta scale times the frequency employed (100 Mcps).

---

*Answers for Page* 9

1. The correspondence between the signals and the components is as follows:

| | |
|---|---|
| ① = II | ⑤ = I |
| ② = IV | ⑥ = III |
| ③ = V | ⑦ = VI |
| ④ = VII | |

2. The cyclohexane peak occurs at 86 cps. If this were the reference signal, the chloroform signal would be at 440 − 86 or 354 cps downfield.

---

*Answers for Page* 11

1. The sharp peak ② is due to the —OH proton.

2. $\overset{④}{CH_3}\overset{③}{(CH_2)_x}\overset{①}{CH_2}\overset{②}{OH}$

3. From the area under the $—CH_2O—$ band (①), it is seen that each integration unit represents 2 protons per 11.1 units or 0.180 proton. The total area under the remainder of the integration curve is 164 units. This corresponds to (0.180) (164) = 29.5 protons. Three of these protons are in the terminal methyl group (band④), and one is in the hydroxyl group (band②). This leaves 25.5 protons for the remainder of the methylene chain. This value, which should be accurate within 10%, corresponds to a compound having the formula $CH_3(CH_2)_{12.8 \pm 1.3}CH_2OH$. The compound used was $CH_3(CH_2)_{12}CH_2OH$. The peak in the integration curve for band ③ is due to the overshooting of the pen.

---

*Answer for Page* 13

1. Although the noise level in this spectrum is not an obvious problem, the ratio of the signal to the noise is still one of the chief limitations of the accuracy of the integration. An improvement in the signal-to-noise ratio could be brought about by: (1) using a more concentrated solution, (2) using a more sensitive instrument (the Varian A-60A has about twice the sensitivity and the HA-100 has about four times the sensitivity of the A-60), (3) repetitive scanning and analyzing the results either with a computer of average transients (C.A.T.) or by simple averaging, or (4) scanning at a slower rate (with the proper radio-frequency power level).

---

## *Answers for Page* 17

1. 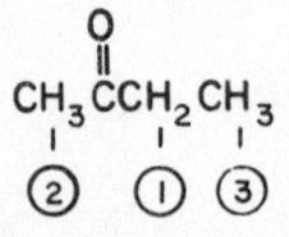 $CH_3\overset{O}{\overset{\|}{C}}CH_2CH_3$ — ② ① ③

2. The two first-order conditions are as follows: (1) the centers of the bands must be separated by at least six times the separation of the peaks in the multiplets, and (2) each proton in one group must be coupled equally to each and every proton in the other group.

3. Yes, this system meets the two first-order conditions. The centers of the bands are separated by 85 cps, while the peaks in the multiplets are separated by 7 cps. There is fast rotation about the bond between the $-CH_2-$ and $-CH_3$ groups. The three stable conformations about this bond are identical. Thus, the angularly dependent vicinal coupling constants between the protons in the $-CH_2-$ and $-CH_3$ groups are averaged and equal.

4. Each part of the pattern consists of $N+1$ peaks, where $N$ is the number of protons in the other group. The adjacent peaks in each band are separated by a distance equal to the coupling constant. The peaks in each multiplet have relative intensities which are approximately equal to the ratios of the coefficients (1:1, 1:2:1, 1:3:3:1, 1:4:6:4:1, etc.) in the expansion of $(R+1)^N$.

5.

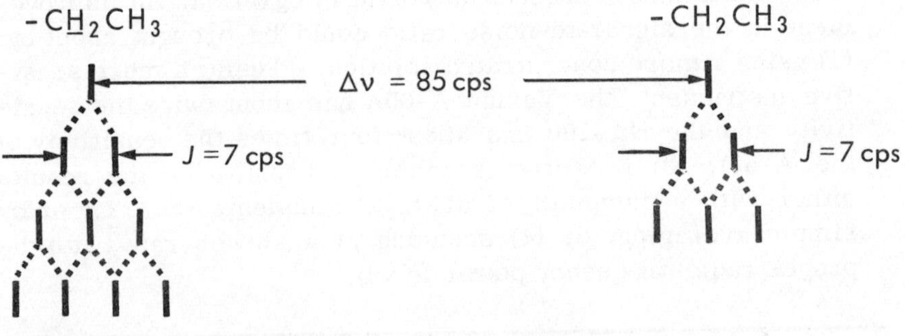

-C<u>H</u>$_2$CH$_3$ -CH$_2$C<u>H</u>$_3$

$\Delta\nu$ = 85 cps

$J$ = 7 cps $J$ = 7 cps

6. The protons in the methyl group are magnetically equivalent. Equivalent protons cannot give rise to multiplets even though they are strongly coupled.

---

*Answers for Page* 19

1. $CHCl_2CH_2Cl$
   ① ②

2. The peaks at 221 cps (band③) and 437 cps ($CHCl_3$) are due to impurities. The peak at 221 cps cannot be part of the first-order multiplet because it is not separated from the other nearest peak by $J$. This peak is undoubtedly due to the presence of $CCl_3CH_2Cl$.

3. Both the rate of rotation about the carbon–carbon bond and the population of the three stable conformations must be considered.

4. This is an $A_2X$ pattern.

5. The relative intensities of the peaks are approximately correct for a first-order pattern. The intensities of the peaks in the triplet should be in the ratio of 1:2:1, while those in the doublet should be in the ratio of 1:1 (or 4:4 if the intensities of these peaks are to be compared with those in the triplet). Note, however, that the peaks of each multiplet which are closer to the other band are slightly taller than the peaks which are furthest removed from the other band. This is one of the three higher-order effects.

---

*Answers for Page* 21

1. $ClCH_2CH_2CH_2Br$
   ① ② ①

2. The signals due to the protons in the $ClCH_2-$ and $-CH_2Br$ groups are split into triplets (2 + 1) by coupling to the middle $-CH_2-$ group. These two triplets both appear in band①. The

signal due to the middle $-CH_2-$ group is split into 2 + 2 + 1 or 5 peaks.

3.

$ClCH_2-$ $-CH_2Br$

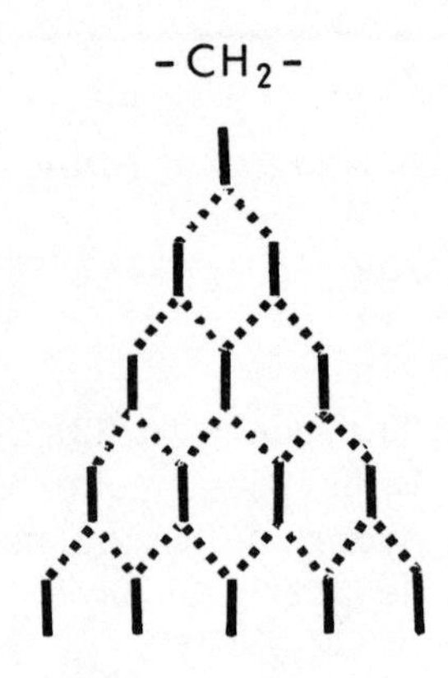


Each $J$ is equal to 6 cps.

---

## *Answers for Page* 23

1.

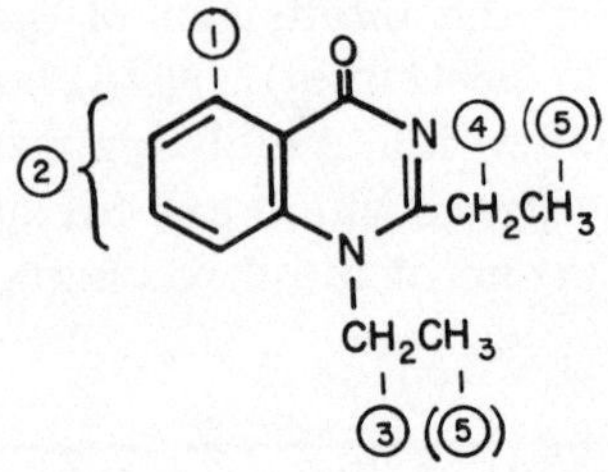

2. The signal due to each of the methylene groups is split by first-order rules to give a quartet (3 + 1) of equally spaced peaks having relative intensities of approximately 1:3:3:1. The methyl protons are split in the same way to give equally spaced triplets (2 + 1) of peaks having relative intensities of 1:2:1. The two triplets can be identified by noting the spacings which must, of necessity, be the same in the quartets and in the corresponding triplets. Peaks a, c, and e constitute one triplet, while peaks b, d, and f constitute the other.

3. The aromatic protons constitute an ABCX system. The multiplet at lowest field is due to the proton (X) *ortho* to the electron-withdrawing carbonyl group. The splittings in the X portion of this higher-order pattern should not be assumed to be equal to coupling constants.

---

## *Answers for Page* 27

1.
```
           CH3-(③)
            |
② - H - COH - ①
            |
           CH3-(③)
```

2. All of the splittings are first-order.

3.

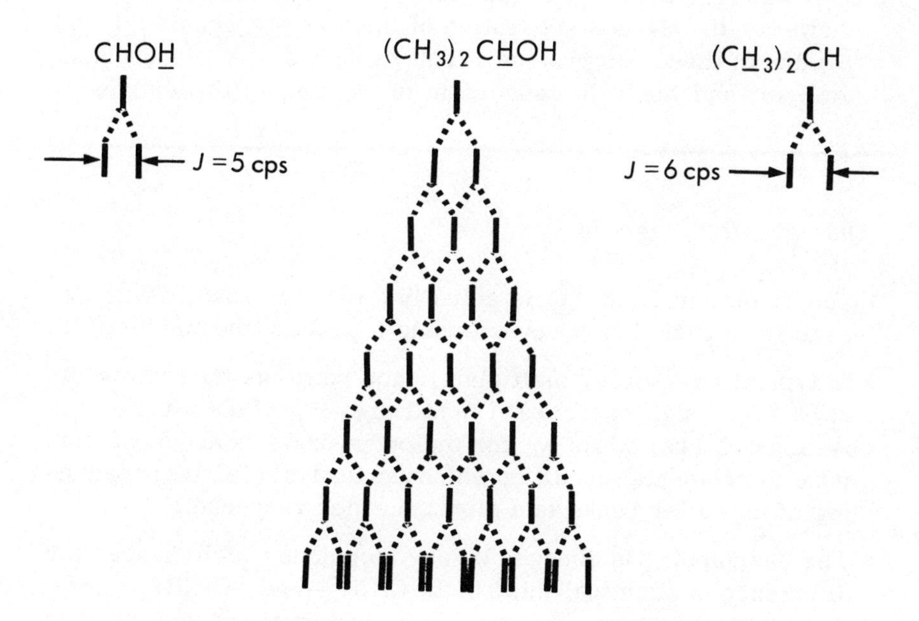


---

## *Answers for Page* 29

$$\begin{array}{l} \qquad\quad CH_3-(③) \qquad\quad ⑤ \\ \qquad\quad | \qquad\qquad\qquad\quad | \\ 1.\ ②-H-COH-④ \quad + \quad HOD \\ \qquad\quad | \\ \qquad\quad CH_3-(③) \end{array}$$

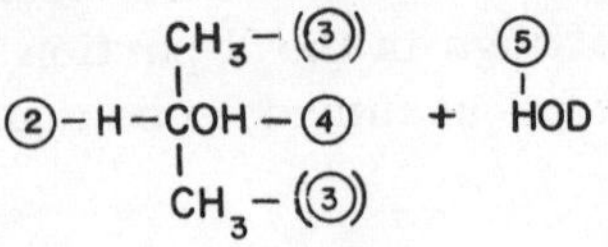

The sum of bands ④ and ⑤ is equivalent to one proton.

2. The addition of $D_2O$ brought about some exchange to give

$$\begin{array}{c} CH_3 \\ | \\ HCOD \\ | \\ CH_3 \end{array}$$

and HOD. The exchange of the proton on the alcoholic hydroxyl group with other protons was taking place at a rate greater than $(\pi/\sqrt{2})(5)$ or 11 times per second. This is revealed by the collapse of the coupling between the HCOH protons, which was observed in the spectrum on p. 26. The exchange between the alcoholic hydroxyl proton and the proton in the HOD was less than $(\pi/\sqrt{2})(34)$ or 75 times per second. This is shown by the 34-cps separation of the distinct signals (④ and ⑤) due to these two protons. Some exchange was taking place, however, and this is the cause of the broadening of these signals.

---

## *Answers for Page* 31

1. The triplet in band ① is actually a pair of doublets with one peak from each doublet overlapping to produce the middle peak.

2. In typical first-order multiplets, each peak has the same width at half the height of the peak ($W_H$). Here, $W_H$ of the middle peak is about 2 cps, while $W_H$ for the outer peaks is about 1.2 cps. (One instrumental condition which can lead to selective broadening of the taller peaks is a slow recorder response.)

3. The overlapping is changed by any technique which changes the difference in chemical shifts of the $ClCH_2$— and —$CH_2Br$ protons. A change of solvent or the use of a higher frequency, such as 100 Mcps, could accomplish this. The best possibilities of changing the difference in chemical shifts by a change in solvent

would be offered by the use of an aromatic solvent, such as benzene (pyridine might cause trouble here), or a solvent having a high dielectric constant, such as deuteroacetone ($CD_3\overset{O}{\overset{\|}{C}}CD_3$).

---

## *Answer for Page* 33

1. 

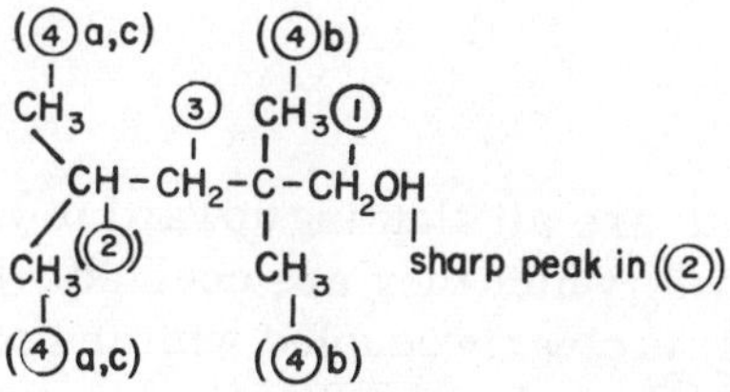

```
 ((4)a,c)      ((4)b)
    |             |
   CH3     (3)  CH3(1)
     \      |     |
      CH-CH2-C-CH2OH
     /  ((2))     |    |
   CH3           CH3   sharp peak in ((2))
    |             |
 ((4)a,c)      ((4)b)
```

---

## *Answers for Page* 35

1. The sharp peak on the low-field side (left) of band (4) is due to the $-NH_2$ protons. This signal is missing after the $D_2O$ exchange.

2. Rapid intermolecular exchange makes it impossible to see the splittings due to such couplings. However, this does not mean that coupling does not occur.

3. 

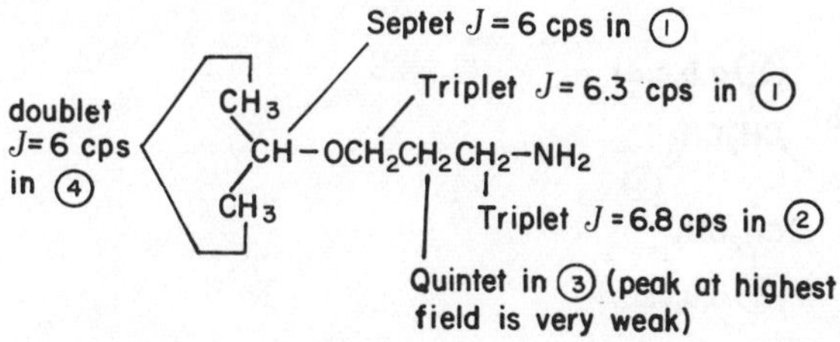

$(CH_3)_2CH{-}OCH_2CH_2CH_2{-}NH_2$

- $(CH_3)_2$: doublet $J = 6$ cps in (4)
- CH: Septet $J = 6$ cps in (1)
- $OCH_2$: Triplet $J = 6.3$ cps in (1)
- middle $CH_2$: Quintet in (3) (peak at highest field is very weak)
- $CH_2{-}NH_2$: Triplet $J = 6.8$ cps in (2)

Of the two outer members of the septet due to the single (methine) proton in $\begin{matrix}CH_3\\ \diagdown\\ \diagup\\ CH_3\end{matrix}CH-$, only the peak at higher field can be seen on this spectrum. This group of peaks is intermixed with the stronger triplet due to the —O—$CH_2$— protons.

4. All of the observed splittings are first-order. The multiplicities are determined by the $N + 1$ rule; the bands are essentially symmetrical about their centers; the peaks in each multiplet are separated by the coupling constant; and the ratio of intensities of the peaks is approximately proportional to the coefficients of the terms in the expansion of $(R + 1)^N$.

---

## *Answers for Page* 37

1. The intensities of the multiplets are all slanting upward toward the signals due to the protons with which they are coupled. For example, the triplet in band (2) is clearly coupled with the protons causing bands (3) or (4) and not with those causing band (1).

2. There could be as many as (2 + 1) (2 + 1) or 9 peaks. The broadening of the peaks in the quintet of band (3) is, in fact, due to the small difference in the coupling constants of the two different groups of methylene protons.

---

## *Answers for Page* 39

1. (H: (1) or (2)), (H: (1) or (2)) C=C (Br) $CH_2$ (3) NH $CH_2$ (5) $CH_2$ ((4) b) N($CH_2CH_3$)$_2$ — $CH_2CH_3$ ((4) a,b,c,d); $CH_3$ (6)

2. The small peak on the high-field side (right) of band (3), the small peak on the low-field side of band (5), and peak (4)e are probably due to impurities. A trace of $CHCl_3$ is also present.

---

## *Answers for Page* 41

1. 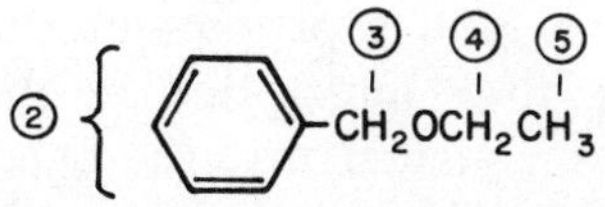 ② { $C_6H_5$–$\underset{③}{CH_2}O\underset{④}{CH_2}\underset{⑤}{CH_3}$

   This spectrum is in general agreement with the structure given. The position of the signal due to the $-CH_2O-$ protons would be expected, from Shoolery's constants, to be at 110 + 142 + 14 or 266 cps. In this spectrum, this signal appears at 270 cps (note the position of the TMS signal). The predicted position for the signal due to the $-OCH_2-$ protons (142 + 28 + 14 or 184 cps) is not in very good agreement with the observed position (212 cps) because neither the electron-withdrawing effect nor the spatial effect of the nearby phenyl group is reflected in the constants which are used. However, these two effects are included in the constants used to predict the position of the signal due to the $-CH_2O-$ protons.

2. The sample is not pure. The aromatic proton region is too complex, and there are extra peaks in bands ③, ④, and ⑤.

   The main impurity is probably $C_6H_5$–$COOCH_2CH_3$. The carbonyl adjacent to the phenyl ring would displace the signals due to the *ortho* protons downfield to give band ①. The signal due to the methylene protons would also be shifted downfield by about 60 cps to give the small peaks in band ③. The signals due to the methyl protons would be displaced downfield slightly to give some of the extra peaks observed in band ⑤.

---

## *Answers for Page* 43

1. The compound used was

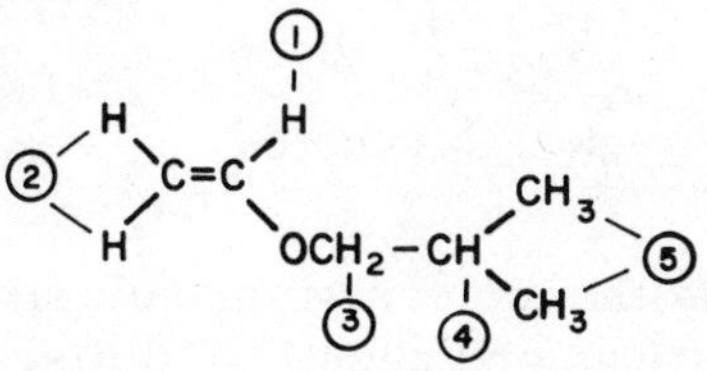

   ② $H_2C=C(H①)-\underset{③}{OCH_2}-\underset{④}{CH}(CH_3)_2$ ⑤

This structure can be deduced in the following manner. The number of protons (12) in the molecular formula is divided by the total integration value (159) to give a proportionality constant. Each band integration is then multiplied by this constant. When these numbers are rounded off, the bands represent the following numbers of protons:

| Band | ① | ② | ③ | ④ | ⑤ |
|---|---|---|---|---|---|
| Protons | 1 | 2 | 2 | 1 | 6 |

The splitting of band ⑤ (6 protons) suggests that it is due to two methyl groups attached to a carbon bearing a single proton $-CH(CH_3)_2$. The single proton would be split into at least 7 peaks and, consequently, must be in band ④. The position and large splittings of the single proton band ① suggest that it is due to the $H_X$ part of a $(H)(H)C{=}C(H_X)-$ system. The signals due to the other two olefinic protons would be in band ②. The small splittings observed within band ② are consistent with the expected geminal coupling. This leaves only the two-proton band ③. These two protons must be next to a carbon bearing a single proton ($-CH_2-\overset{|}{C}H-$). The groups arrived at up to this point are: $(H)(H)C{=}C(H)-$, $-CH_2-\overset{|}{C}H-$, and $-CH(CH_3)_2$.

There are only 12 protons in the compound, so that the carbon atom which bears the single proton must be placed adjacent to both the methylene and the two methyl groups to give $-CH_2CH(CH_3)_2$. The oxygen atom can now be used to join the two groups to give

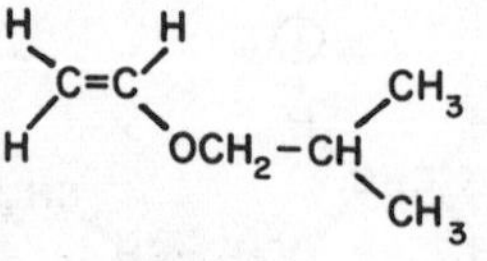

$$H_2C{=}CH-OCH_2-CH(CH_3)_2$$

This structure fits the molecular formula and is consistent with the observed splittings and chemical shifts.

2. The olefinic protons constitute an ABX system. Note that the electron-withdrawing oxygen atom causes a downfield shift of the signal due to the proton attached to the same carbon atom. However, the contribution of the polarized form ($\overset{\ominus}{C}H_2{-}CH{=}\overset{\oplus}{O}{-}$) causes an increase in electron density at the terminal carbon atom $\left(\begin{matrix} H \\ H \end{matrix}\!\!>\!C{=}\right)$. Thus, the signals due to the protons attached to this carbon atom are shifted to relatively high fields.

---

*Answers for Page* 45

1. 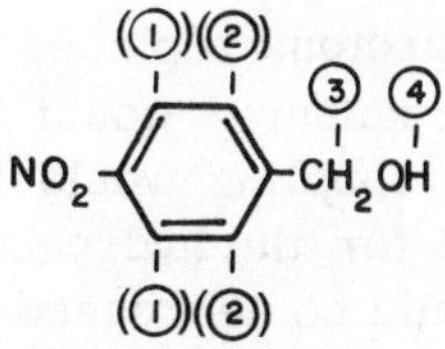 $NO_2$–$C_6H_4$–$CH_2OH$ (p-nitrobenzyl alcohol; ring protons labeled ((1))((2)) on each side, $CH_2$ labeled (3), OH labeled (4))

2. In the first determination, there was not enough acid present in the $CDCl_3$ to catalyze the rapid intermolecular exchange of the hydroxyl proton. In the spectrum on p. 44, the first-order coupling of the hydroxyl proton and the methylene protons ($J = 5$ cps) gives rise to the doublet (3) and the triplet (4). With a small amount of hydrochloric acid present, the intermolecular exchange rate became much faster than ($\pi/\sqrt{2}$) (5) or 11 exchanges per second. The coupling, although it still exists, can then no longer be observed in the spectrum on p. 46.

3. The relative intensities of the peaks are distorted in the "wrong" direction. The tallest peak should be on the side toward band (4). This result was brought about by the use of too much radio-frequency power ("saturation").

---

*Answers for Page* 47

1. A medicine dropper that contains vapor from the space above the liquid in a concentrated hydrochloric acid bottle could be

introduced over the solution in the NMR tube. The hydrochloric acid added to the solution in this manner would increase the exchange rate even more. The increased exchange rate would cause the bands to sharpen.

2. The aromatic protons constitute an $A'_2B'_2$ or AA'BB' system.

---

## *Answers for Page* 49

1. The slight broadening of the high-field side of the $A'_2B'_2$ pattern is probably due to a small coupling with the $-CH_2-$ protons.

2. The compound having the two methylene protons replaced by deuterium could be synthesized. The broadening would be reduced for this compound, because the coupling would be decreased to $^1/_{6.55}$ of the value observed for the hydrogen-containing compound. The broadening would be decreased by only $^2/_{6.55}$, however, because each of the peaks would now be split into five peaks $(2N+1)$ rather than into three $(N+1)$. The $A'_2B'_2$ pattern should then be essentially symmetrical. Another method would be to irradiate band ③ with a strong signal and observe band ②. This "double-resonance" experiment would decouple the aromatic protons and the methylene protons.

---

## *Answers for Page* 51

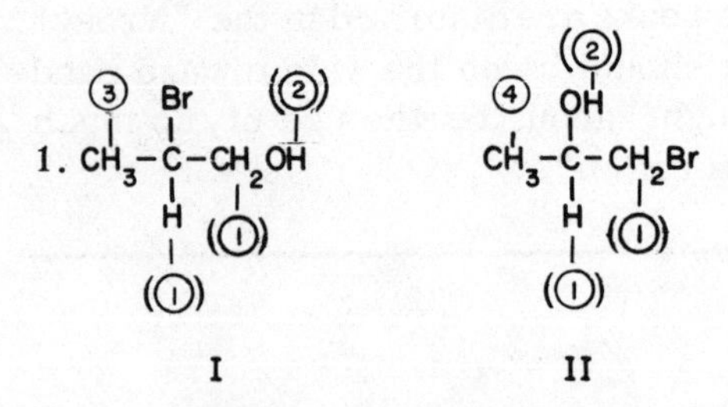


2. The mixture consists of about 25 and 75 mol.% of I and II, respectively. These percentages can be calculated in various

ways. For example, the integration of band (3) can be divided by the sum of bands (3) and (4). This gives the molar ratio of I in the mixture. Another way is as follows: The total integration represents 7 protons, so that each unit represents 7/200.5, or 0.035 proton. Band (3) integrates for (0.035) (20.5 units), or 0.716 proton. This corresponds to 0.716/3, or 0.24 of a methyl group, or 24% of I.

---

*Answer for Page* 53

1. The compound used was

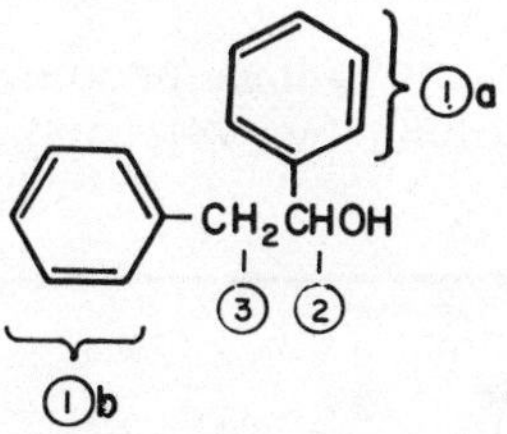

$$C_6H_5CH_2CHOH(C_6H_5)$$

The aromatic proton signals are assigned by comparison with the signals due to the aromatic protons in toluene (430 cps) and benzyl alcohol (436 cps). Sharp singlets are fairly characteristic of the signal due to aromatic protons in compounds of

the type $C_6H_5\text{—C—}$.

---

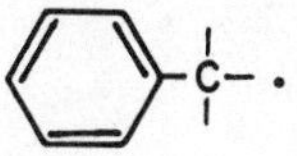

*Answers for Page* 55

1.

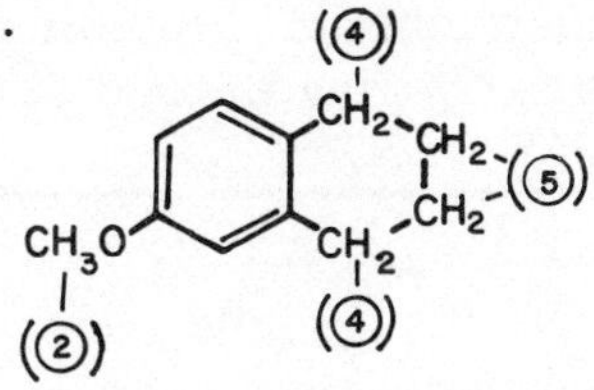

I

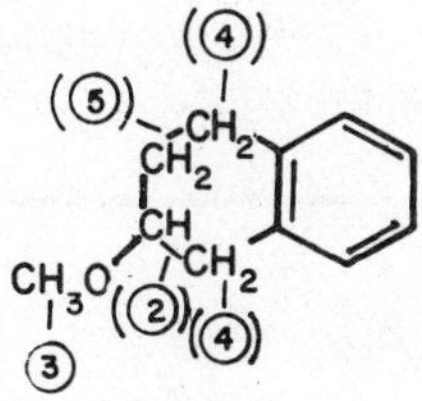

II

The proton on the carbon bearing the methoxy group in II is probably a very broad band at the base of the tallest methoxy peak (②). Band ① contains the signals due to the four nearly equivalent aromatic protons in II (tallest peak) and the non-equivalent aromatic protons in I (an ABC system).

2. The mixture consists of approximately 75 and 25 mol.% of I and II, respectively. Each unit of integration represents 14/222, or 0.063 proton. The small amount of $CHCl_3$ is ignored here. Band ③, which represents the three protons in the aliphatic methoxy group, amounts to (13.5) (0.063), or 0.851 proton. This corresponds to 100 (0.851/3), or 28 mol.% of II. Band ② represents not only the three protons in the aromatic methoxy group, but also the broad band due to the ring $CH_3OC\underline{H}$ proton. This band should amount to 0.28 proton. The mole percentage of I should thus be given by $\frac{1}{3}$ [(40.5) (0.063) − 0.28] 100, or 76 mol.%.

---

*Answers for Page* 57

1.

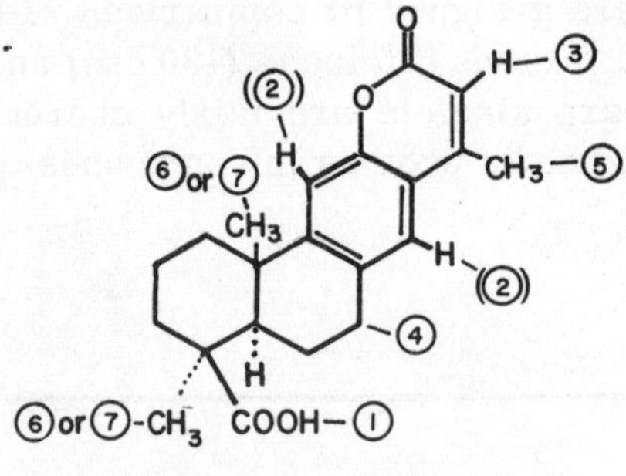

2. Any $CHCl_3$ in the $CDCl_3$ will show up in band ②. Differences in the local magnetic field experienced by the protons in solvents are very effectively averaged out by free tumbling. This often causes strong ringing. This may explain the ringing in band ②.

---

*Answers for Page* 59

1. The HOD peak is at 276 cps.

2. The slight elevation in the integration curve near 220 cps is probably not due to contamination with methanol because the methanol would probably be washed out in the $D_2O$ exchange. The peak is very likely due to contamination with the corresponding methyl ester ($-COOCH_3$).

---

## *Answers for Page* 61

1.

$OCH_3$–(3)
(5) or (6) b
$CH_3$
(1)
[(4)]
[(4) b,d,f,g]
[(4) a,c,e]
$CH_2CH_3$
$CH_3$ $COOCH_2CH_2N$
(6) a,c,d
$CH_2CH_3$
(5) or (6) b
(2)
[(4) b,d,f,g]

Band (4) corresponds to 8.7 protons. It includes the six protons on the carbon-bearing nitrogen and the two protons on the carbon next to the phenyl ring. (See the spectrum on p. 56.)

2. The aromatic protons constitute an ABC system which could have as many as 15 peaks.

---

## *Answers for Page* 63

1.

(2)
(5)
H
$CH_3$
$CH_3$
C=C
(4)
$CH_3$
$CH_3$
H
H
$CH_3$
(5)
$COOCH_2CH_3$
(3)(5)
I

(5)
$CH_3$
H
$CH_3$
C=C
(4)
$CH_3$
H
(1)–H
$CH_3$
$CH_3$
(5)
$COOCH_2CH_3$
(3)(5)
II

From the molecular model, the olefinic proton on the same side as the carbonyl (structure II) should be assumed to be further downfield (more deshielded) than the olefinic proton on the opposite side of the carbonyl (structure I). The protons on the cyclopropane ring are deshielded relative to the signal due to the protons in cyclopropane itself.

2. A change of solvent might cause a change in the chemical shifts involved in band ⑤, which would be helpful in sorting out the peaks.

3. The mixture consists of about 58 and 42 mol.% of I and II, respectively. This can be determined from the integrations of the olefinic proton bands. The sum of the two bands is 9.5 units. The component with the olefinic proton signal at lower field (II) amounts to 4 units. This means that II is present to the extent of (4/9.5) (100), or 42%. Because of the probable error in the measurement of these small integrations, this estimate is not expected to be very accurate.

   Essentially the same result should be obtained by use of the relative heights of the peaks in the $-CH_2-$ band. The more intense quartet must be due to the methylene protons in I.

---

*Answers for Page* 65

1.

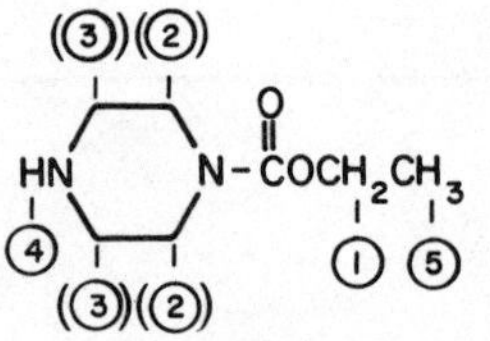

(③)(②)

HN ⟨ring⟩ N-COCH$_2$CH$_3$

④ ① ⑤

(③)(②)

The $\underline{H}N-$ proton signal is removed by the $D_2O$ exchange. The $-OC\underline{H}_2CH_3$ protons are downfield because of the electron-withdrawing effect of the oxygen. The ring methylene protons closest to the carbonyl are further downfield because of the electron-withdrawing effect of the carbonyl and the average spatial relationship between these protons and the carbonyl.

2. Bands ① and ⑤ are parts of a first-order pattern. Bands ② and ③ are parts of a higher-order pattern. The single peak ④

arises from the HN– proton, which is undoubtedly undergoing fast intermolecular exchange.

3. The $-OCH_2CH_3$ group constitutes an $A_3X_2$ system, and the ring methylene protons constitute two superimposed $A'_2X'_2$ systems. Another way of designating the ring protons is AA'XX'.

---

## *Answer for Page* 67

1. The direction of the distortion of a multiplet indicates whether the other part of the pattern is at a higher or a lower magnetic field, while the degree of distortion indicates the distance to the other part of the pattern. Note, for example, that, even if the splittings were equal, band ③ and band ⑤ could not be related because they are slanted in the same direction. Bands ① and ③ are slanted toward each other, but band ③ is distorted more than band ①.

---

## *Answers for Page* 71

1. ①,④ O ②,③

$HP(OCH_3)_2$

The peak at 449 cps and part of the peak at 196 cps are due to impurities. The peak at 196 cps coincides with a small spinning side band.

2. Couplings to phosphorus obey the same rules as couplings to hydrogen. The single proton is coupled through one bond to the phosphorus to give the doublet ① and ④, which is separated by $J_{H\text{-}P}$ (695 cps). The methoxy protons are coupled through three bonds to the phosphorus to give the doublet ② and ③ with $J_{\underline{P}\text{-}OC\underline{H}_3}$ equal to 12 cps.

3. Since the H–P coupling is first-order, the chemical shift of the proton is halfway between the two members (① and ④) of

the doublet. The TMS signal is 2 cps downfield, so that $\nu_H$ is equal to $\frac{1}{2}(59 + 754) - 2$, or approximately 405 cps downfield from TMS at 60 Mcps.

---

## *Answers for Page* 73

1. Peaks ③, ④, ⑤, and ⑥ in each scan are spinning side bands. These spinning side bands are easily identified because of the dependence of their positions on the sample spinning rate.
2. The sample was spinning faster during the determination of scan B.
3. Peak ② is due to the protons in water, which was present as an impurity. Peaks ① and ⑦ are $C^{13}$ satellite signals. The value for coupling between $C^{13}$ and the protons attached to this carbon atom is 137 cps. The two methyl groups in this compound are equivalent. The natural abundance of $C^{13}$ is 1.1%. This means that approximately (1.1) (2) or 2.2% of the molecules contain one $C^{13}$ atom. The intensity of each of the two $C^{13}$ satellite signals is, therefore, approximately 1.1% of the intensity of the main peak.

---

## *Answers for Page* 75

1. ①,③,④ → $CHF_2$; ((②)) → COOH; ② → HOD

$$CHF_2COOH + HOD$$

The $\overset{|}{\underset{|}{\text{C}}}$H— proton is split ($J = 53.5$ cps) by the two fluorine atoms according to first-order rules to give (2 + 1) or 3 peaks. The small peaks at the base of band ② are spinning side bands. One spinning side band causes the shoulder on the low-field side of band ③.

2. Dilution with $D_2O$ would be expected to change the amount of hydrogen bonding of the carboxylic acid proton and also to shift

the averaged signal in the direction of the expected position of the HOD signal.

---

## *Answers for Page* 79

1. ① – $\overset{\overset{O}{\|}}{HC}N(\overset{③}{CH_2}\overset{④}{CH_3})_2$

   Band ② is due to an impurity.

2. The slow rotation about the C–N bond caused by partial double-bond character makes the two ethyl groups nonequivalent.

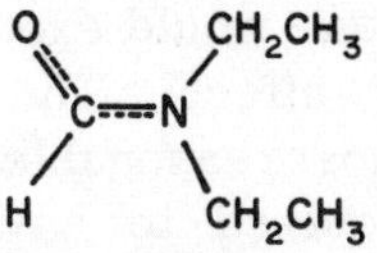

$$\begin{array}{ccc} O & & CH_2CH_3 \\ & C{=\!=}N & \\ H & & CH_2CH_3 \end{array}$$

   Both ethyl groups give the expected first-order triplet–quartet patterns. These patterns are overlapping. In band ③, peaks a, c, e, and g belong to one quartet, while peaks b, d, f, and h belong to the other. In band ④, a, c, and e constitute one triplet, and b, d, and f constitute the other.

---

## *Answers for Page* 81

1. $\overset{O}{\|}$ HCNHCH$_2$CH$_3$ (H ①, NH ②, CH$_2$ ③, CH$_3$ ④)

2. The signal due to the –NH– proton is broadened by the following three factors: (1) coupling with the nitrogen nucleus, which should give a 1:1:1 triplet having separations of approximately 50 cps; (2) the nonspherical distribution of charge on the nitrogen nucleus, together with the unsymmetrical electrical environment of the nitrogen nucleus; and (3) small couplings with the $\overset{O}{\overset{\|}{HC}}$– and –$CH_2$– protons.

3. The methylene protons are coupled both to the $-CH_3$ protons and to the –NH– proton with essentially the same (7 cps) coupling constant. This first-order splitting thus gives 3 + 1 + 1, or 5 peaks.

4. The three most reasonable explanations for these spikes are as follows: (1) further coupling, (2) nonequivalence due to restricted rotation about the C–N bond, and (3) presence of an impurity. Explanation (1) appears unlikely because the splitting would appear to be due to coupling with a single proton. The intensities are distorted in the wrong direction to indicate coupling with the –NH– or $H\overset{\overset{O}{\|}}{C}-$ protons. If the spikes were due to nonequivalence, one would expect the $-CH_2-$ band ③ to show an even greater effect. The presence of an impurity would seem to be the most reasonable cause of the spikes. The impurity would be required to have an ethyl group almost identical to the ethyl group in this compound. The presence of $H\overset{\overset{O}{\|}}{C}N(CH_2CH_3)_2$ is a possibility.

5. Slow interconversion between the two stable conformations about the C–N bond could cause the appearance of two nonequivalent $-CH_2CH_3$ patterns.

---

*Answers for Page* 83

1. After the $D_2O$ exchange, the methylene proton signal is split primarily by coupling with only the methyl protons. The coupling with the –NH– proton is now replaced by a coupling with –ND–, which is only $^1/_{6.55}$ as large as the coupling with –NH–. The coupling thus results in a quartet of closely spaced ($J = 1$ cps) triplets ($2N + 1$ modification of the $N + 1$ rule) having relative intensities of 1:1:1. The fine splitting cannot be seen here.

2. The broadness of band ① is probably due in part to coupling with the nitrogen nucleus and in part to the nonspherical charge distribution on the nitrogen nucleus.

---

*Answers for Page* 85

1. Band ③ is due to HOD.

2. Because the exchange was incomplete, band ④ is due to the overlapping of the quintet expected before exchange (see p. 80) and the quartet expected after exchange (see p. 82).

---

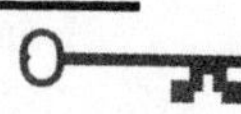

*Answers for Page* 87

1.

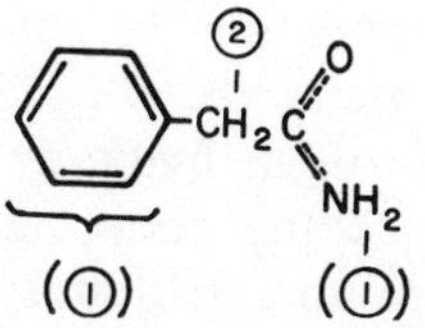

The two $-NH_2$ protons are nonequivalent because of the partial double-bond character of the $-C-N-$ bond. The broad signals due to these two protons appear at the base of band ①. The integration of amide protons is commonly low. Note the low ratio of the integrations of bands ① and ②. The lettered peaks are spinning side bands which occur about 24 cps on each side of the sharp peaks in bands ① and ②. The quintet in band ③ is due to residual $CHD_2\overset{O}{\overset{\|}{S}}CD_3$ in the solvent.

2. The phenyl protons are essentially equivalent. Equivalent protons cannot give rise to observable multiplicity even though they are strongly coupled.

---

*Answers for Page* 89

1.

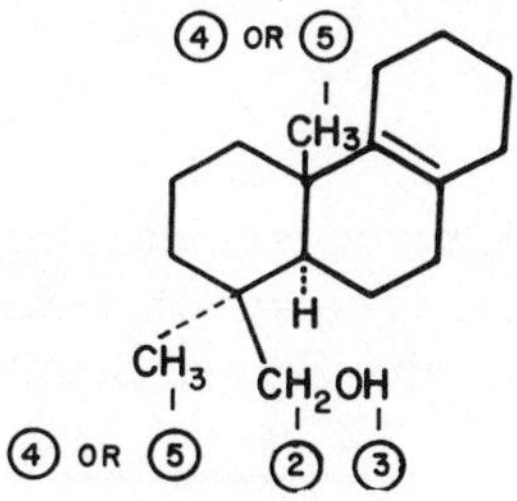

2. Structure II must be correct because the only olefinic proton signal present is band ①, which is of very low intensity. Band ① is probably due to the presence of a very small amount of either I or III.

3. The protons in the $-CH_2O-$ group constitute an AB system. The nonequivalence of these two methylene protons is a consequence of the three different substituents on the adjacent carbon atom. The nonequivalence of these methylene protons next to a carbon bearing three different substituents is not necessarily due to restricted rotation.

4. The 11-cps coupling constant is consistent for the coupling between methylene protons on a carbon bearing an hydroxyl group. The presence of an hydroxyl group or halogen atom lowers the magnitude of the coupling constant from 12.4 cps (actually, −12.4 cps) to approximately 10 cps. The magnitude of the coupling constant would be increased by approximately 1.9 cps by the presence of an adjacent double bond and by approximately 3.8 cps by an adjacent triple bond.

5. Because of steric hindrance, it is probable that the spatial arrangement of the hydrogen atoms and hydroxyl group in the $-CH_2OH$ group are relatively fixed. In this fixed orientation, there is a stereospecific long-range coupling between one of the $-CH_2-$ protons and one or more of the nearby ring protons.

---

*Answers for Page* 93

1.

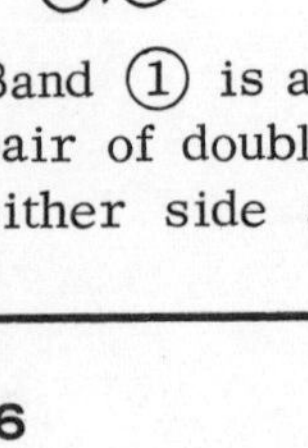

$CH_3$–②
H, H ①
$O_2$
CH
$CH_3$ $CH_3$
③, ④

2. Band ① is an AB pattern. The central, stronger peaks of the pair of doublets are very close together. The small peaks on either side are the outer members of the pair of doublets.

---

*Answers for Page* 95

1. The parameters are related as follows:

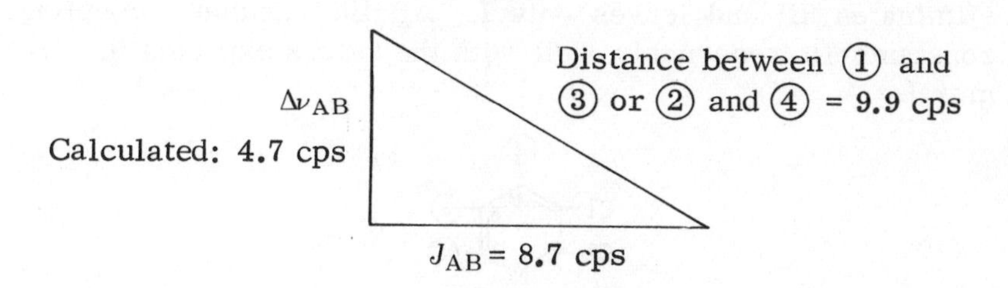


The center of the AB pattern is at ½(387.9 + 386.7), or 387.3 cps. The chemical shift of $H_A$ is thus 387.3 − ½(4.7), or 384.9 cps; the chemical shift of $H_B$ is 387.3 + ½(4.7), or 389.7 cps.

2. If a first-order treatment were employed, the chemical shifts of A and B would be taken as the midpoints of the doublets. Thus, the two chemical shifts would be ½(396.6 + 387.9) or 392.3 and ½(378.0 + 386.7) or 382.3 cps. Note that the actual difference in chemical shifts is 4.7 cps, rather than 10 cps as approximated from the first-order treatment.

3. The expected intensity ratio of band ① to band ② (and band ④ to band ③) would be (387.9 − 386.7)/(396.6 − 378.0), or 1.2/18.6, or 0.065. The observed average is ½(8 + 7)/½(139), or 7.5/69.5, or 0.11.

---

*Answers for Page* 97

1. This pattern is due to an ABX system. The three protons are clearly different from each other. The pattern approaches an AMX, but bands ② and ③ are separated by a value only four times the estimated coupling constant between them.

2. Isomers IV and VI would, because of symmetry, have a maximum of only two types of protons.

3. The coupling constants are approximately 1.5, 4.7, and 8.0 cps.

4. The largest coupling constant of 8.0 can only be between protons located at C-3 and C-4. This eliminates isomers II and V and leaves only I and III to be considered. The second largest coupling constant, 4.7, is too large for coupling between a proton at C-2 and a proton at either C-4 or C-5. This eliminates III and leaves only I. All the estimated coupling constants fit reasonably well with the values expected in isomer I.

5.

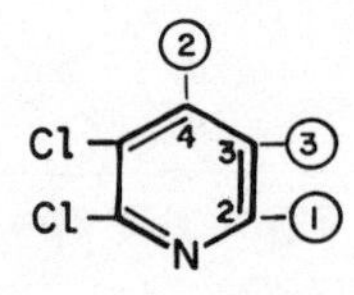


These assignments can be made by matching the observed and expected splittings. The C-4 proton should be coupled strongly with the C-3 proton and weakly with the C-2 proton. Band (2) shows such splittings. In the same way, the C-3 proton should be coupled strongly with the C-4 proton and moderately with the C-1 proton. Band (3) shows splittings which fit these requirements. The remaining band, band (1), must be due to the C-2 proton. This band shows the expected moderate coupling to the C-3 proton and weak coupling to the C-4 proton. The signals due to the $\alpha$-protons on pyridine are usually further downfield than the signals due to the other protons on the ring.

---

*Answers for Page* 99

1. These are $A_2B$ patterns.

2. The compound used has structure IV. Because of the large downfield shift of the $\alpha$-protons on a pyridine ring, isomer VI would be expected to give an $A_2X$ pattern rather than an $AB_2$. (See "Answers for Page 97.")

3. Scan A was determined on a Varian HA-100 instrument by Ross G. Pitcher of Varian Associates. Scan B was determined on a Varian A-60 instrument. Note that in scan A there is a greater separation of the A and B portions of the pattern and

smaller higher-order splittings of both the central member of the triplet and the higher-field member of the doublet.

4. Scan B approximately matches pattern 2-4 in the text by Wiberg and Nist[1] ($\Delta\nu_{AB}/J_{AB}=3$) and the pattern for $\Delta\nu_{AB}/J_{AB}=4$ in the article by Corio.[2] In scan B, the separation between peak c and the mean position of peaks e and g is 21.8 cps. This is equal to $\Delta\nu_{AB}$. Thus, $J_{AB}$ must be about 5.5 to 7 cps. At 100 Mcps, $\Delta\nu_{AB}$ should be (100/60) (21), or 36.4 cps. The value found was 35.8 cps.

   If $\Delta\nu_{AB}/J_{AB}$ were reduced further, the tallest peaks (labeled e and f) would split into doublets. For the measurements used in the following discussion, it is assumed that these two peaks exactly coincide.

   A simple way[3] of extracting $J_{AB}$ from an $AB_2$ pattern is to use the following formula:

$$J_{AB} = \tfrac{1}{3}\,(a-d + f-h)$$

where a–d and f–h refer to the distances between the peaks as they are lettered in the spectrum. In this example, then,

$$J_{AB} = \tfrac{1}{3}\,(15 + 9) = 8 \text{ cps}$$

   The value of $J_{AB}$ can also be extracted[3] from the pattern of the B nucleus alone by using the following formula:

$$J_{AB} = \tfrac{1}{3}\,[e-g + 2(f-h)]$$

In this example,

$$J_{AB} = \tfrac{1}{3}\,[6.5 + 2(9)] = 8.2 \text{ cps}$$

---

[1] K. B. Wiberg and B. J. Nist, The Interpretation of NMR Spectra, W. A. Benjamin, Inc. (New York), 1962, p. 15.
[2] See p. 381 of the article by P. L. Corio. Chem. Rev. 60:363 (1960).
[3] These methods are taken from a lecture by Dr. Aksel Bothner-By presented at the Fourth Workshop on High Resolution Proton Magnetic Resonance held at the College of Pharmacy of the University of Illinois, Chicago, Illinois, December 15-17, 1965.

*Answer for Page* 101

1. Patterns A, D, and E are ABC types. Pattern B is an ABX, while C is an AMX. The compounds used were

A — $H_A$, $H_C$, C=C, $H_B$, $SO_2CH_3$

D — $CH_3O$, H, H, H, N, H

B — O, $H_A$, C—C, $H_X$, $H_B$

E — H, H, C=C, H, $[\;]_2$ $Si(CH_3)_2$

C — $H_A$, $H_X$, C=C, $H_M$, $OCCH_3$, O

Note that C, A, and E form a sequence of vinyl-proton patterns in which the major variation is the difference in chemical shifts between X and the AB protons.

Note also that the distortion of the peaks from equal intensities in the multiplets can be used to judge the nature of the system. All of the peaks in the AMX (pattern C) are essentially of equal height. The peaks in the AB portion of the ABX (pattern B) are distorted to some extent, while all of the peaks in A and especially in E are considerably distorted.

*Answers for Page* 103

1.

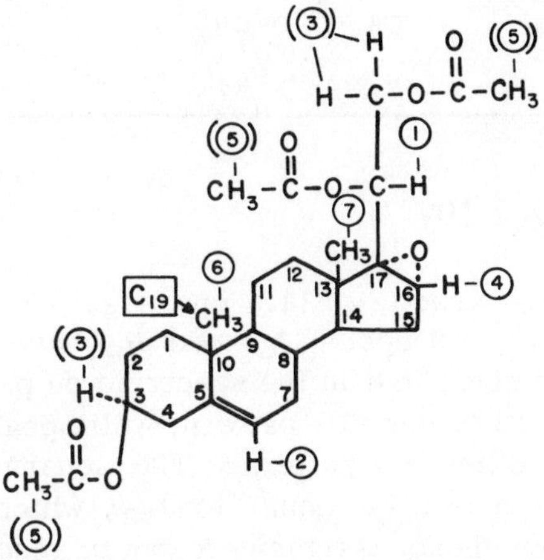

The broad band on the low-field side of band ③ is due to the C-3 proton. The additive constants for the substituents on the D-ring, which are required for the calculation of the angular methyl positions, are not given by Zürcher.[1] The position of the C-19 methyl signal can be estimated, however, by assuming only small contributions for the D-ring substituents. The constants for the C-3 $\beta$ acetoxy group (+3.0 cps) and the C-5,6 double bond (+14.0 cps) are thus added to the base value of 47.5 cps. This calculated value (64.5 cps) is in better agreement with band ⑥ than with band ⑦.

2. Band ④, which is due to the C-16 proton, is sharp because of the very small coupling of this proton with the protons on the adjacent carbon atom. Coupling constants having unexpectedly large or small magnitudes constitute a potential pitfall in the interpretation of spectra.

3. The sharp peaks in bands ① and ③ belong to an ABX system.

[1] R. F. Zürcher, Helv. Chim. Acta 46:2054 (1963). Zürcher's data are also presented and discussed in the text by Bhacca and Williams, Chapter 2.

These signals are due to the $H_AC(H_B)(-O-\overset{O}{\overset{\|}{C}}-CH_3)-CH_X(-O-\overset{O}{\overset{\|}{C}}-CH_3)$ protons.

---

## *Answer for Page* 107

1. The parameters are: $|J_{AB}| = 11.5$ cps; $|J_{AX}| = 8.8$ cps; $|J_{BX}| = 3.7$ cps; and $\Delta\nu_{AB} = 8.9$ cps. ($J_{AX}$ and $J_{BX}$ have the same sign.)

   First, it is noted that in the spectrum on p. 104 peaks a, c, d, and g constitute one AB pattern, while peaks b, e, f, and h constitute the other AB pattern. The separations a–c, d–g, b–e, and f–h are each equal to $|J_{AB}|$, which is 11.5 cps. A cross check on these assignments can be made by noting that the centers (245.8 and 239.8 cps) of the two AB patterns are separated by 6.0 cps. This value should be equal to one-half the separation ($|J_{AX} + J_{BX}|$) of the two strongest peaks in the X pattern. Peaks a and d on p. 106 are separated by 12.5 cps. A further check can be made by noting that the separation (2.8 cps) of peaks b and c in the X pattern (p. 106) is approximately equal to the difference (3.1 cps) between the two distances, a–d (13.1 cps) and b–f (16.2 cps), in the AB portion (p. 104). The two remaining peaks in the X portion should be separated by the sum (29.3 cps) of the two distances, a–d and b–f. These two peaks, which are not observed, must have very low intensities. The parameters are related as follows:

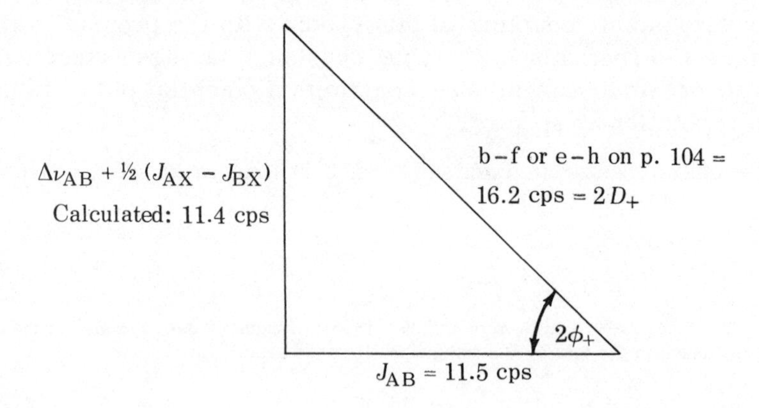

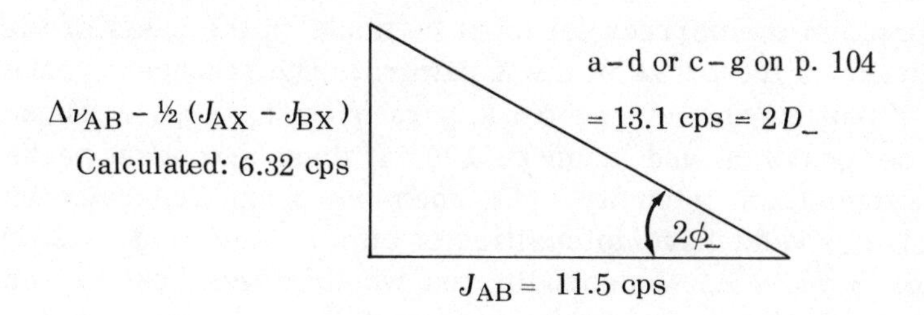


It follows from the triangles above that

$$\Delta\nu_{AB} + \tfrac{1}{2}(J_{AX} - J_{BX}) = \pm\ 11.4$$

and

$$\Delta\nu_{AB} - \tfrac{1}{2}(J_{AX} - J_{BX}) = \pm\ 6.32$$

Subtraction yields

$$|J_{AX} - J_{BX}| = 5.1 \text{ or } 17.7 \text{ cps}$$

It is known that $|J_{AX} + J_{BX}| = 12.5$ cps, since this is the separation of the strong outer peaks of the X pattern. It follows then that

$$|J_{AX}| = \tfrac{1}{2}(5.1 + 12.5) = 8.8 \text{ cps}$$

or

$$\tfrac{1}{2}(17.7 + 12.5) = 15.1 \text{ cps}$$

This means that

$$|J_{BX}| = 3.7 \text{ cps}$$

or

$$-2.6 \text{ cps}$$

Addition of the two expressions for the vertical sides of the triangles gives

$$2|\Delta\nu_{AB}| = 6.32 \pm 11.4 \text{ cps}$$

Thus

$$|\Delta\nu_{AB}| = 8.9 \text{ or } 2.6 \text{ cps}$$

The two possible sets of values, then, are

$$|J_{AB}| = 11.5;\ |J_{AX}| = 8.8;\ |J_{BX}| = 3.7;\ |\Delta\nu_{AB}| = 8.9$$

$$|J_{AB}| = 11.5;\ |J_{AX}| = 15.1;\ |J_{BX}| = -2.6;\ |\Delta\nu_{AB}| = 2.6$$

The choice of the correct set must be made on the basis of the intensities of the peaks in the X pattern. The strongest peaks in the X pattern should be those separated by $|J_{AX}+J_{BX}|$. These must be peaks a and d on p. 106. If these strongest peaks are assigned an intensity of 1, then the peaks separated by $(2D_+ - 2D_-)$ should have intensities of either $\cos^2 \frac{1}{2}(2\phi_+ - 2\phi_-)$ or $\cos^2 \frac{1}{2}(2\phi_+ + 2\phi_-)$, depending on whether the first or the second set of the above values is correct. In the same way, the intensities of the weakest peaks should be either $\sin^2 \frac{1}{2}(2\phi_+ - 2\phi_-)$ or $\sin^2 \frac{1}{2}(2\phi_+ + 2\phi_-)$. It can be seen that

$$\cos 2\phi_+ = 11.5/16.2$$

and

$$\cos 2\phi_- = 11.5/13.1$$

Thus,

$$2\phi_+ = 44.8^\circ$$

and

$$2\phi_- = 28.5^\circ$$

This means that

$$\cos^2 \tfrac{1}{2}(2\phi_+ - 2\phi_-) = \cos^2\left(\frac{16.3^\circ}{2}\right) = \cos^2 8.1^\circ = 0.98$$

and

$$\cos^2 \tfrac{1}{2}(2\phi_+ + 2\phi_-) = \cos^2\left(\frac{73.2^\circ}{2}\right) = \cos^2 36.7^\circ = 0.642$$

The nearly equal intensities of the four peaks in the X pattern mean that the triangles are related as

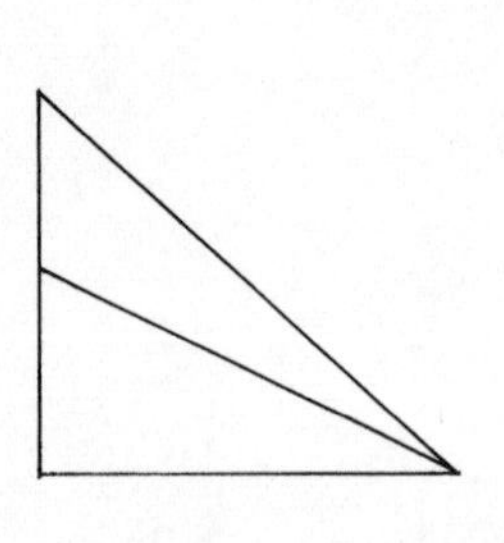

rather than as

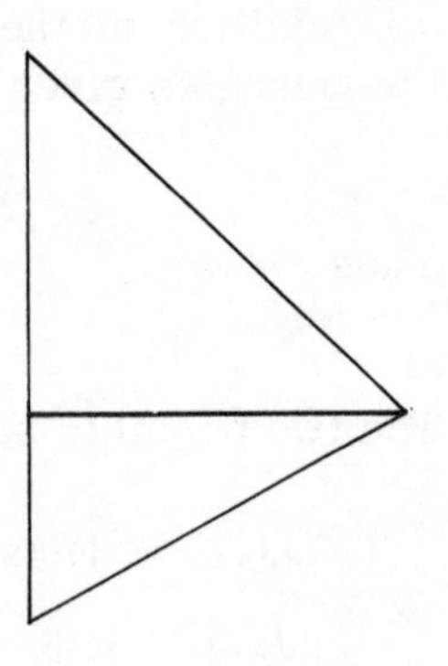

Correspondingly, the weakest peaks in the X pattern should have (1−0.98) or 0.02 times the intensities of the strongest peaks. These peaks are lost in the noise.

---

## *Answers for Page* 109

1. The phenyl protons probably constitute an $AB_2C_2$ system. The other three protons constitute an ABX system, which is approximately an AMX.

2. $C_6H_5$–$CH_B$=$CH_A$–C(=O)$H_X$  (phenyl ring: (2); $H_B$: (2) a,b; $H_A$: (3); $H_X$: (1))  $J_{AX}$ = 7.2 cps, $J_{BX} \sim 0$; $J_{AB}$ = 16 cps

3. Structure I must be correct. The value of $J_{AB}$ is too large for the cis coupling which would be observed for II.

4.

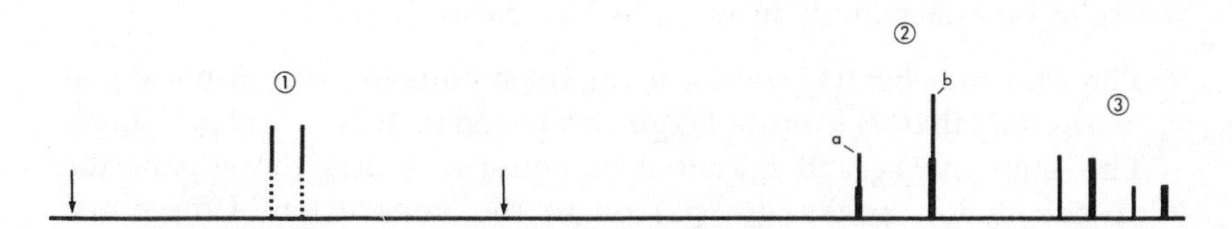


Peaks a and b in band ② constitute a pair of superimposed doublets, each of which is identical to the doublets (alternating peaks) in band ③. The weakest peaks in the X portion would be $2D_- + 2D_+$ or 98 cps apart. The other four peaks are all essentially of equal intensity and are superimposed in pairs to give the doublet in band ①.

---

## *Answers for Page* 111

1. No, this compound probably is not pure. Note that the signals due to the two groups of angular methyl protons are not of the

same height, even though they are approximately of the same width at half the peak heights. Band (2) is probably due to the C-19 methyl protons in the impurity. These conclusions are tentative because it is possible that peaks and valleys in the underlying absorption curve are responsible for the differences in the heights of the peaks.

2\. $H_B$ $H_A$ $CH_3$ $H_X$ Br O H

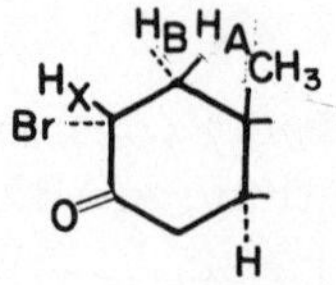

Band (1) is the X portion of either an AMX or an ABX pattern. The two protons at C-1 are expected to be nonequivalent because the bond to one of these protons (shown dotted) is essentially perpendicular to the plane of the ring (axial), while the bond to the other proton is essentially in the plane of the ring (equatorial).

In a number of similar substances in which a complete analysis was possible,[1] this type of system approximated an AMX pattern.

3\. No, if this were the X portion of an ABX system, the splittings would be controlled, in part, by the ratio $\Delta\nu_{AB}/J_{AB}$.

4\. The fact that band (1) is not a triplet means that $J_{AX}$ is not equal to $J_{BX}$ and that $J_{AB}$ is not large compared to $\Delta\nu_{AB} + \frac{1}{2}(J_{AX} - J_{BX})$. The sum of $J_{AX}$ and $J_{BX}$ must be equal to either the separation of the inner peaks (6 cps) or to the separation of the outer peaks (20 cps) in band (1). (It can be said from experience that the sum of an axial–axial coupling and axial–equatorial coupling should be larger than 6 cps.)

---

## *Answers for Page* 113

1\. Each pattern is symmetrical about the center of the total pattern. Pattern C is approximately symmetrical about the center of each half of the pattern.

[1] R. J. Abraham and J. S. E. Holker, J. Chem. Soc. 806 (1963). These cases, together with another example, are discussed by Bhacca and Williams on p. 144 of their text.

2. The patterns are as follows: (a) *p*-disubstituted phenyl, C and G; (b) aliphatic $-CH_2-CH_2-$ group, A and F; (c) flexible six-membered ring containing two $-CH_2-CH_2-$ groups, E; and (d) symmetrical *o*-disubstituted phenyl, D. B and D appear much the same, but the position of B precludes this as a pattern due to an aromatic system. The compounds used were

A $ICH_2CH_2COOH$

B $CH_2-O$, $CH_2-O$, S, O

C I–(phenyl)–$OCH_3$

D (quinoxaline) N, N

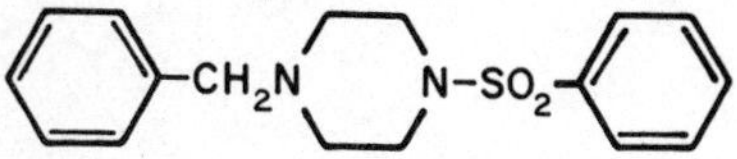

E (phenyl)–$CH_2N$ (piperazine) N–$SO_2$–(phenyl)

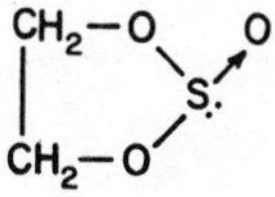

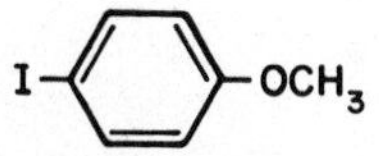

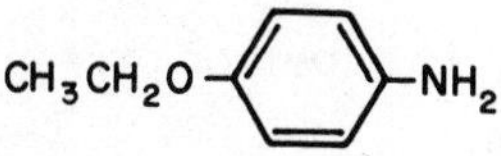

F $NCCH_2CH_2SCH_2CH_2CN$

G $CH_3CH_2O$–(phenyl)–$NH_2$

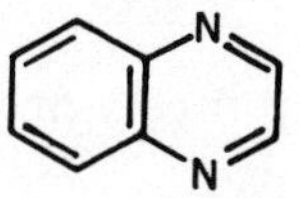

3. Patterns A and F are approximately of the $A_2B_2$ type. Pattern C is approximately an $A'_2X'_2$. All of the other patterns are $A'_2B'_2$ types.

Note the slight difference in the manner in which the outer peaks in A and F have broken up. Of the two patterns, A appears to be somewhat more consistent with calculated $A_2B_2$ patterns.[1,2] The differences in these patterns may be due to unequal populations of the three stable conformations about the $-CH_2-CH_2-$ bonds. With unequal populations, two different coupling constants between the groups may be involved. This, in turn, makes the pattern dependent, in part, on the geminal couplings between the protons in the methylene groups.

---

[1] K. B. Wiberg and B. J. Nist, The Interpretation of NMR Spectra, W. A. Benjamin (New York), 1962, p. 320.

[2] See p. 395 of the article by P. L. Corio, Chem. Rev. 60:363 (1960).

## *Answers for Page* 117

1. The compound used was

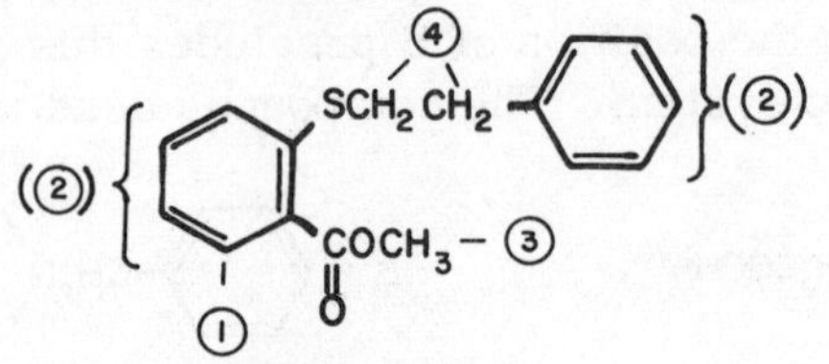

First, it is noted that the number of protons represented by each of the bands is as follows:

| Band | ① | ② | ③ | ④ |
|---|---|---|---|---|
| Number of protons | 1 | 8 | 3 | 4 |

Tentatively, the sharp signal in band ② can be assigned to the protons in a $C_6H_5$– group, while the sharp three-proton signal in band ③ can be assigned to the protons in a $-OCH_3$ group. The downfield position of the one aromatic-proton signal in band ① suggests the presence of a second phenyl group having a substitution pattern such as 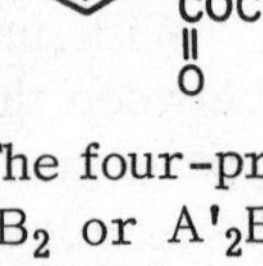 . It seems reasonable to attach the $-OCH_3$ to give 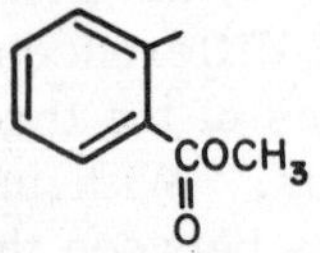 . At this point the remaining atoms are $C_2H_4S$. The four-proton signal in band ④ appears to be due to an $A_2B_2$ or $A'_2B'_2$ system. The ratio $\Delta\nu/J_{AB}$ is so small that much of the character of this band is lost. The ratio of the elements suggests that the grouping might be $-SCH_2CH_2-$. This fragment can be used to join the two phenyl groups to give

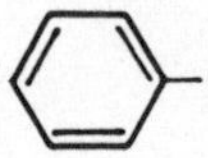

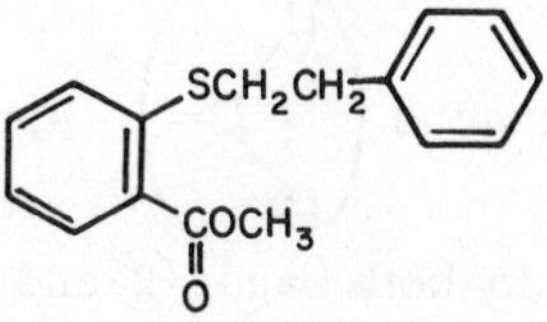

This is in agreement with the molecular formula. The small difference in chemical shifts in the $-CH_2CH_2-$ group is consistent with the difference expected on the basis of Shoolery's constants for the substituents (98 and 110 cps or 1.64 and 1.83 ppm).

2. The following systems are present: $A_3$ ($-OCH_3$); $A_5$ (phenyl); $A_2B_2$ (or possibly an $A'_2B'_2$, $-SCH_2CH_2-$); and an ABCX (*o*-disubstituted phenyl).

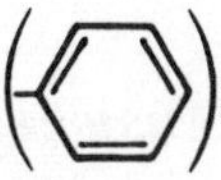

## *Answers for Page* 119

1. 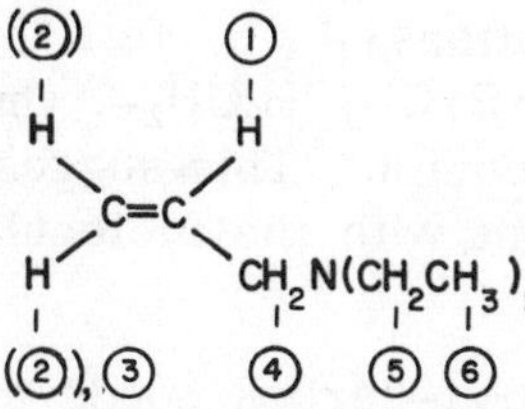 $H$ (②) and $H$ ①, $C{=}C$, $H$ (②), ③ and $CH_2N(CH_2CH_3)_2$ with ④, ⑤, ⑥

2. Two identical $A_3X_2$ systems and an $A_2XYZ$ system are involved.

3. The mutual coupling ($J \approx 7.5$ cps) of the methyl and methylene protons in the ethyl groups gives the first-order pattern seen in bands ⑤ and ⑥. The methylene proton signal in band ④ is split into a doublet ($J \approx 6.0$ cps) by the adjacent olefinic proton. Each member of the doublet is further split by couplings ($J \approx 1$ cps) to the other two olefinic protons. The olefinic proton signal in band ① is split by different couplings ($J_{trans} \approx 17$ cps and $J_{cis} \approx 9$ cps) to the other two olefinic protons and to the methylene protons ($J \approx 1$ cps). The signal due to one of the

geminal olefinic protons $\left( \begin{matrix} H \\ \quad \diagdown C= \\ H \diagup \end{matrix} \right)$ appears in band ②, while the other appears in both bands ② and ③. These two protons are coupled weakly with each other and with the methylene protons ($-CH_2-$) and are strongly coupled with the other olefinic $\left( \diagdown C{=}C\diagup^{H} \right)$ proton.

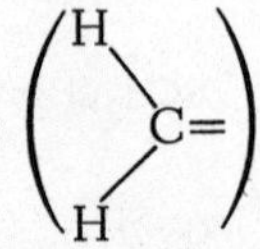

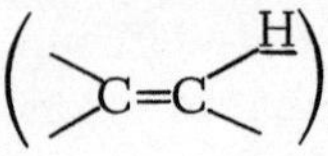

---

*Answers for Page* 121

1. The compound used was $BrCH_2CH_2CH_2CN$. This structure can be deduced as follows: Band ① represents two protons, while band ② represents four protons. The two protons represented by band ① must be coupled to two adjacent protons. It seems reasonable to postulate the grouping $BrCH_2CH_2-$. The two remaining protons in band ② are also coupled with other protons, otherwise there would be a sharp singlet ($A_2$) or at most a pair of doublets (AB) with equal separations ( $J_{AB}$ ). This then suggests the extension of the grouping to $BrCH_2CH_2CH_2-$. Only one carbon and one nitrogen atom remain. This suggests $BrCH_2CH_2CH_2CN$, which is in agreement with the molecular formula.

2. Assuming fast rotation about the carbon–carbon bonds and equal conformations, the spin system would be designated as an $A_2B_2X_2$.

---

*Answers for Page* 123

1. The compound used was

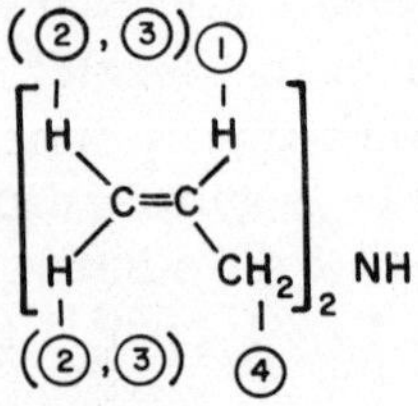

$$\left[ \begin{matrix} \underset{(②,③)}{H} & & \underset{①}{H} \\ & C{=}C & \\ \underset{(②,③)}{H} & & \underset{④}{CH_2} \end{matrix} \right]_2 NH$$

2. A spectrum determined at 100 Mcps would be approximately first-order. A different solvent might be found which would also cause sufficient changes in the chemical shifts to give a first-order spectrum.

---

*Answers for Page* 125

1. The compound used was

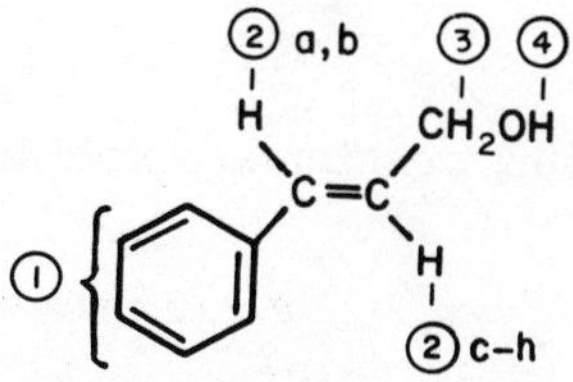

2. Bands (2) and (3) constitute an $A_2XY$ (or an $ABX_2$) pattern. Both members of the high-field doublet of the XY pattern are split into triplets. The smallest peak (f) of the triplet (f, g, h) is lost in the noise. If $J_{XY}$ were much larger, higher-order splittings would occur in band (3).

   The coupling (15 cps) between the two olefinic protons is equal to the separations in band (2) between peaks a and b, d and g, and e and h. The large magnitude of this coupling proves that the compound has the trans rather than the cis configuration.

---

*Answers for Page* 127

1.

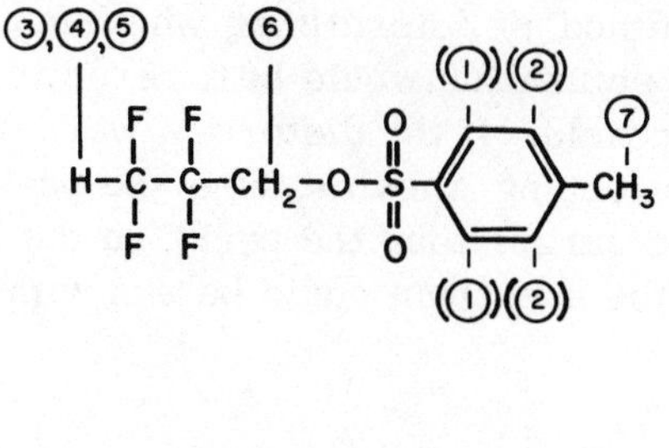

Fluorine has a spin quantum number of ½. Thus, the same multiplicity rules hold for F–H couplings as for H–H couplings. The very large couplings which may occur between F–F and between F–H increase the number of problems that may arise because of higher-order effects.

2. Band ② is probably broadened because of coupling with the protons in the $-CH_3$ group. The assignment of the aromatic protons is made on this assumption and on the downfield shift expected to result from the electron-withdrawing effect of the $-O\overset{O}{\underset{O}{\overset{\|}{\underset{\|}{S}}}}-$ group.

3. The two H–F coupling constants which may be measured with assurance are

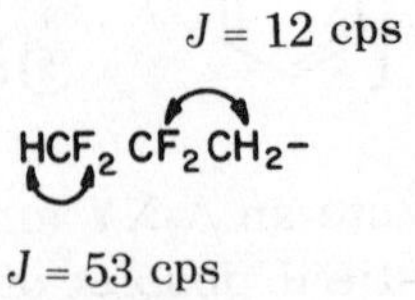

$J$ = 12 cps

$HCF_2CF_2CH_2-$

$J$ = 53 cps

The finer splittings in bands ③ – ⑥ must not be taken as coupling constants, because these splittings could be the result of "virtual" coupling. The fluorine spectrum would have to be analyzed before it could be determined whether or not the systems involved are first-order. Until a complete analysis was made, the finer splittings could be referred to as "apparent" coupling constants or simply cited as splittings.

4. The fluorine nuclei absorb too far upfield to cause this much distortion in the intensities of the multiplets. Furthermore, the distortion is in the wrong direction. The distortion is probably a result of saturation (too much RF power). Rapid scanning of the spectrum can also cause this type of distortion.

5. The easiest method of determining whether or not the distortion was due to saturation would be to rerun the spectrum from higher to lower field. If the distortion were due to saturation, the reversed sweeping should cause the peaks on the higher-field side to be larger than the peaks on the lower-field side. Alternatively, the spectrum could be run with lower RF power

to determine if less distortion occurs. A slower sweep speed would correct the distortion if it were caused by too rapid scanning.

---

## *Answers for Page* 129

1. The spectrum on p. 128 corresponds to II; the one on p. 130 corresponds to I; and the spectrum on p. 132 corresponds to III.
2. None of the separations in this higher-order pattern (band ③) may be taken as coupling constants.

---

## *Answers for Page* 131

1. It is first noted that the band assignments are

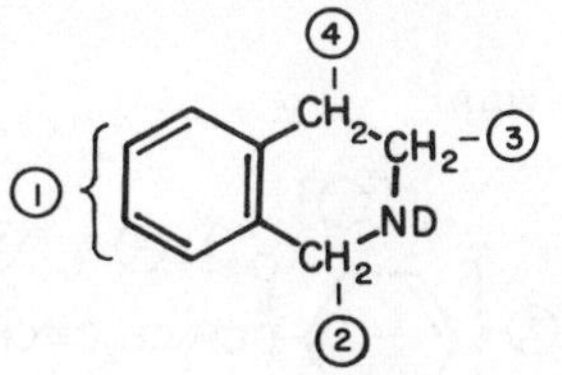

   The assignments of bands ③ and ④ are made by noting that the lower-field band (left) on p. 132 (in band ③) is of lower intensity than the higher-field band. It is reasonable to assume the same relative positions of these bands on the spectrum on p. 130.

   The two most probable causes of the broadening of band ④ are: (1) coupling to the closest aromatic proton, or (2) a slow rate of interconversion between different conformations of the ring.
2. If the broadening were due to further coupling, band ④ could be sharpened by either double resonance or by replacement of the aromatic proton with deuterium. If the broadening were

due to slow interconversion between the ring conformations, an elevation of the temperature of the solution should cause band ④ to sharpen.

---

## *Answers for Page* 133

1. The spectrum could be run at 100 Mcps, which would increase all of the differences in chemical shifts by $^{100}/_{60}$. A change in solvent might also cause an increase in the differences in chemical shifts. Spin decoupling of the methyl protons (band ④) by double resonance could be used to remove the extra coupling to band ③, reducing it essentially to an ABC pattern.

2. These peaks could be due to the presence of a small amount of the compound used on p. 132.

---

## *Answers for Page* 135

1. The compound used was

[①] (ortho), ② (meta, para) $C_6H_5-\overset{O}{\overset{\|}{C}}-\overset{③}{CH_2}\overset{④}{CH_2}C\equiv\overset{⑤}{CH}$

The reasoning which could lead to this structure is as follows: First, the integrations indicate that the numbers of protons represented by the bands are

| Band | ① | ② | ③ | ④ | ⑤ |
|---|---|---|---|---|---|
| Number of protons | 2 | 3 | 2 | 2 | 1 |

The total number of protons assigned (10) is in agreement with the molecular formula. By the positions and splittings in bands ① and ②, these signals appear to be due to the five protons in a mono-substituted phenyl group. The total number of "rings" (double bonds count as two-membered rings) in the compound must be equal to C + 1 $-\frac{1}{2}$H or 11 + 1 $-\frac{1}{2}(10) = 7$. Four of these "rings" are in the phenyl group, leaving three for the side chain. None of the remaining bands (③, ④, and ⑤) are far enough downfield to be due to olefinic proton signals. The position and small splittings of band ⑤ (one proton) suggest that this could be due to an acetylenic proton in a grouping such as $-CH_2C{\equiv}CH$. The signal due to the methylene protons would also have to show this same small splitting. Band ④ (two protons) appears to meet this requirement. The protons in band ④ are also coupled with other protons to give approximately a triplet. These protons must be represented by band ③. Bands ③, ④, and ⑤ could thus be due to the protons in the grouping $-CH_2CH_2C{\equiv}CH$. The only remaining unused atoms are C and O. The position of band ③ suggests that it is next to an electronegative group. The structure

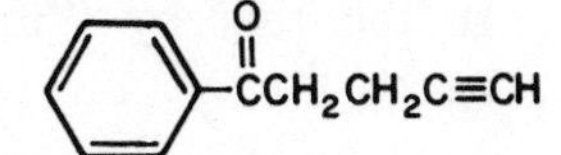


would fit all of these observations. The chain contains the required additional three "rings." The two protons represented by band ① are the phenyl protons *ortho* to the carbonyl group. The remaining phenyl protons are represented by band ②.

2. The aromatic protons constitute an $A'B'_2C'_2$ system (further separation of bands ① and ② would be required before this could be classified as an $A'B'_2X'_2$). The remaining protons might, at first consideration, be expected to constitute an $AX_2Y_2$ or approximately an $AM_2X_2$ system. It will be noted, however, that band ③ does not have the appearance of half of a typical $A_2B_2$ pattern (compare with pattern A on p. 112, or the calculated patterns on pp. 318-323 in the text by Wiberg and Nist[1] or on p. 395 of the article by Corio[2] in Chemical Reviews). In typical $A_2B_2$ patterns, the outer peak of the triplet splits at a larger ratio of $\Delta\nu/J$ than the central peak. Clearly, no "virtual" coupling is involved because the separations of bands ③, ④, and ⑤ are reasonably large compared to the corresponding coupling constants between the protons. The deviation of the

[1] K. B. Wiberg and B. J. Nist, The Interpretation of NMR Spectra, W. A. Benjamin, Inc. (New York), 1962.
[2] P. L. Corio, Chem. Rev. 60:363 (1960).

pattern in band ③ from calculated patterns for $A_2B_2$ systems must be a consequence of either slow rotation about the bond between the methylene groups or unequal populations of the three stable conformations about this bond. The spin system involved must correspondingly be an $A'X'_2Y'_2$ or approximately an $A'M'_2X'_2$ type.

3. The acetylenic proton is split into a triplet by a first-order coupling with the two closest methylene protons. The two groups of methylene protons constitute a higher-order ($A'_2B'_2$, or approximately an $A'_2X'_2$) system. Each peak in one half (band ④) of this $A'_2B'_2$ pattern is split further into doublets by the first-order coupling to the acetylenic proton.

4. In the absence of coupling between the protons represented by bands ④ and ⑤, band ④ would be the mirror image of band ③. The two halves of the $A'_2B'_2$ pattern on p. 327 (Figure 5-20) in the text by Wiberg and Nist have an appearance similar to band ③.

---

## *Answers for Page* 137

1. Band ④ was due to residual $CHD_2\overset{O}{\overset{\|}{S}}CD_3$ in the solvent. This should be a quintet. [$2NI + 1$. $I$ for deuterium is equal to unity; $N = 2$. Thus, $2(2)(1) + 1 = 5$.]

2. The sample used was $CH_3\overset{O}{\overset{\|}{C}}NH-C_6H_4-COOH$. The following discussion is a reasoning process by which this structure can be deduced. Examination of the integration data indicates that the bands are due to the following numbers of protons:

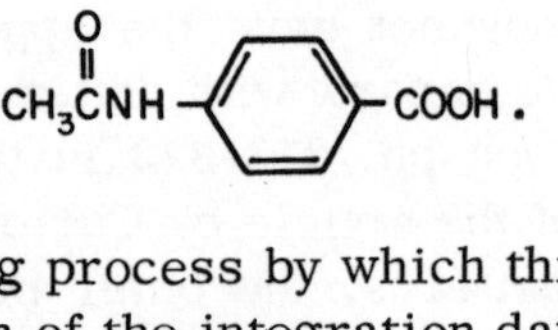

| Band | ① | ② | ③ | ⑤ |
|---|---|---|---|---|
| Number of protons | 1 | 1 | 4 | 3 |

This checks with the number of protons in the molecular formula. Bands (1) and (2) are removed by the $D_2O$ exchange. The broadness of band (1) is consistent with a secondary-amide proton ($-\overset{O}{\overset{\|}{C}}NH-$). Band (2) will be assigned later by elimination. The position and appearance of the four-proton band (3) suggest that this is due to a *p*-disubstituted phenyl proton. Band (5) must be due to three equivalent, or nearly equivalent, protons. This suggests a $CH_3-$ group. If the nitrogen is assumed to be trivalent, there must be $C + 1 - \frac{1}{2}(H - N)$ or $9 + 1 - \frac{1}{2}(9 - 1)$ or 6 "rings" in the molecule. The phenyl ring counts as four "rings" (three double bonds and one six-membered ring), and the amide counts as one "ring." This leaves one "ring" unassigned, which can be easily accommodated in a $-COOH$ group. This would give the other required exchangeable proton (peak (2)).

At this point, the following groups have been suggested: $CH_3-$, $-\overset{O}{\overset{\|}{C}}NH-$, $-COOH$, and $-C_6H_4-$ (p-phenylene). The methyl group can be located on any of the other groups, except $-COOH$, to give

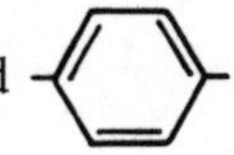

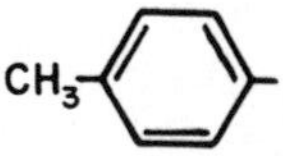

$$CH_3-C_6H_4- \qquad CH_3\overset{O}{\overset{\|}{C}}NH- \quad \text{or} \quad -\overset{O}{\overset{\|}{C}}NHCH_3$$

I II III

Grouping III can be ruled out by the chemical shift which is expected to be near 160 cps rather than 126 cps. This position fits II better than I. In addition, a small coupling between the methyl protons in I and the two *o*-phenyl protons would be anticipated, which would broaden half of the $A'_2B'_2$ system. With the assumption that grouping II fits best, the structure must be

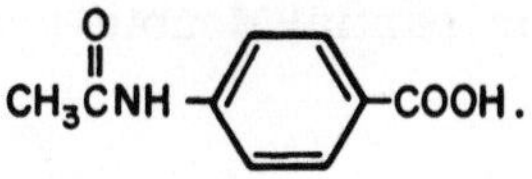

$$CH_3\overset{O}{\overset{\|}{C}}NH-C_6H_4-COOH.$$

## *Answers for Page* 139

1. The dilution of the sample with $D_2O$ caused exchange of the $-\overset{O}{\overset{\|}{C}}NH-$ and $-COOH$ protons, giving the single sharp HOD peak at 254 cps (band ⑥).

2. The rate of exchange between the $-\overset{O}{\overset{\|}{C}}NH-$ and $-COOH$ protons in the solution used for the spectrum shown on p. 136 was much less than $(\pi/\sqrt{2})$ (140) or 310 exchanges per second. The two signals would coalesce to give a single broad signal when the exchange rate was of the same order of magnitude as $(\pi/\sqrt{2})$ times the value for the separation of the two signals in the absence of exchange. Very rapid exchange would give rise to a single sharp peak.

---

## *Answers for Page* 141

1. $C_6H_5-\overset{②}{CH_2}N(C_6H_5)\overset{O}{\overset{\|}{C}}\overset{③}{NH}\overset{④}{CH_2CH_2}Cl$ (both phenyl groups: ①)

2. In the absence of further coupling, the four protons in the $-CH_2CH_2Cl$ group would be expected to give approximately an $A_2X_2$, $A_2B_2$, or $A_4$ pattern. All of these patterns would be symmetrical about the midpoints of the patterns. The lack of symmetry could be due either to coupling between the protons on the adjacent nitrogen and carbon atoms ($-NH-CH_2-$) or to nonequivalence resulting from the partial double-bond character of the N–C bond.

---

# *Answers for Page* 145

1.

$$\begin{array}{c} \text{(1)((2))} \\ \text{S}{=}\text{C}(\text{CH}_3)\text{-N}<\begin{smallmatrix}\text{CH}_2\text{CH}_2 \\ \text{CH}_2\text{-CH}_2\end{smallmatrix}>\text{O} \\ \text{(3)} \quad \text{((2))((2))} \end{array}$$

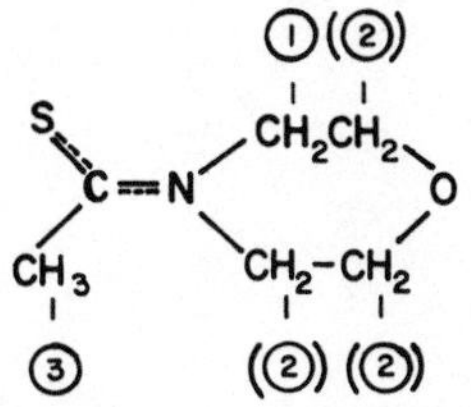

Restricted rotation due to the partial double-bond character of the C–N bond causes the $-N\langle^{CH_2-}$ groups to have different spatial relationships to the $\overset{S}{\overset{\|}{-C-}}$ group. This, in turn, makes the $-N\langle^{CH_2-}$ groups magnetically nonequivalent. The protons in one of the $-CH_2CH_2-$ groups appear as an $A'_2B'_2$ type, where half of the pattern appears as band (1) and the other half as part of band (2). The protons in the other $-CH_2CH_2-$ group are almost equivalent, and even though they are undoubtedly coupled just like the corresponding protons in the other $-CH_2CH_2-$ group, equivalent protons cannot interact in such a way as to give observable multiplicities. The result is essentially a single peak due to an $A_4$ pattern.

2. The rate of rotation about the C–N bond would be increased; therefore, the effect of the $\overset{S}{\overset{\|}{-C-}}$ group would be the same for both $>N-CH_2-$ groups. This would cause both of the $>N-CH_2-$ groups to become identical. The result would be two identical superimposed $A'_2B'_2$ systems with an averaged $\Delta\nu_{AB}$. Each half of the pattern would look roughly like band (1), except that there would be a greater distortion in the relative peak intensities because of the smaller separation of the bands.

## *Answers for Page* 147

1. (structure: $(CH_3)_2N$–C₆H₄–NO with protons labeled ③ on $(CH_3)_2N$, ② and ① on ring positions)

2. Because of the slow rotation about the bond between the ring and the nitroso group, the compound behaves as if it were a mixture of the following two identical forms:

$H_A$ $H_X$ / $(CH_3)_2N$–ring–N=O / $H_{A'}$ $H_{X'}$      $H_{A'}$ $H_{X'}$ / $(CH_3)_2N$–ring–N=O / $H_A$ $H_X$

The protons labeled $H_A$ and $H_X$ (same side as the oxygen atom) are displaced downfield compared to the protons labeled $H_{A'}$ and $H_{X'}$ (opposite the oxygen atom). With a very slow rate of interconversion between the forms, this system would be an ABXY or approximately an $A_2XY$ system. With fast interconversion, the system would become an $A'_2X'_2$. At an intermediate rate, some averaging would take place so that the system would appear intermediate between the $A_2XY$ and the $A'_2X'_2$ systems. Since the *ortho* protons are affected the most, the pattern resembles an aromatic $A'_2X'_2$ with a broadened $X_2$ portion.

---

## *Answers for Page* 149

1. (structure: ring $CH_2$–$CH_2$ ① bonded through O, O to $CH_3$–C–$CH_2CH_3$, with $CH_3$ ③, $CH_2$ ②, $CH_3$ ④)

2. The protons constitute $A_3$, $A_4$, and $A_3X_2$ systems.

3. This is a first-order pattern. The signal due to each group of protons is split because of the protons on the adjacent carbon atom. The number of peaks is determined by the $N + 1$ rule. The methyl signal is thus split into 3 + 1 peaks. The methylene signal is split into 2 + 1 peaks.

4. The protons within each of these two groups are equivalent. Equivalent protons cannot give rise to observable splitting even though they are strongly coupled with each other.

---

## *Answers for Page* 151

1.

2. The $A'_2B'_2$ band (2) is due to the four protons in the ketal group. The methylene protons are nonequivalent because of the difference in their spatial relationship with the phenyl and methyl groups. The protons on adjacent carbon atoms which are on the same side of the five-membered ring are equivalent to each other. A conventional drawing often obscures the stereochemistry. The corresponding ketal band in the spectrum on p. 148 is an $A_4$ pattern because the methyl and ethyl groups are not sufficiently different to cause nonequivalence.

3. Only about 3 mol.% of the starting material remains. This value is obtained by a comparison of bands (3) and (4). All of the molecules must have a methyl signal either in band (3) or in (4). The molar percentage of starting material is thus (100) (1.8)/(1.8 + 54.5) or 3.2 mol.%.

---

## *Answers for Page* 153

1.

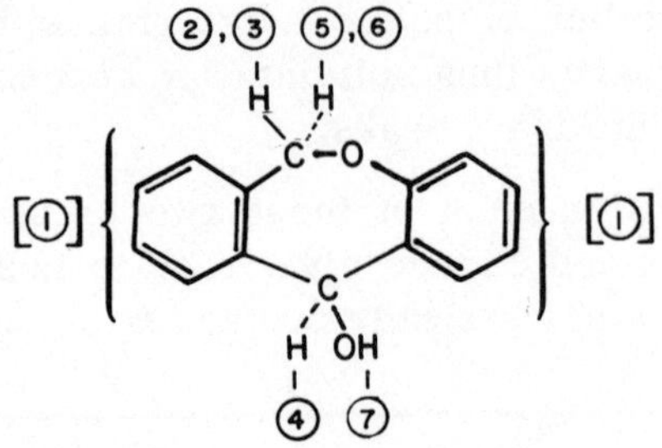

The methylene protons are nonequivalent because they have different spatial relationships with the –OH and –H in the

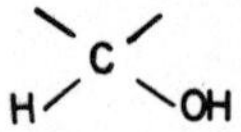

group.

2. The –OH proton is coupled with the proton on the same carbon. The intermolecular exchange rate is greater than $(\pi/\sqrt{2})(5)$ or 11 exchanges per second, but is not rapid enough to cause complete collapse of the smeared doublets into sharp peaks.

3. Addition of a trace of hydrochloric acid or other strong acid would increase the exchange rate, causing peaks ④ and ⑦ to sharpen.

---

## *Answer for Page* 155

1. The preparation of the acetate would shift peak ④ downfield by about 60 cps (1 ppm) and would probably change the difference in chemical shifts between the two methylene protons. No appreciable change in the geminal coupling constant is to be expected.

---

## *Answers for Page* 157

1.

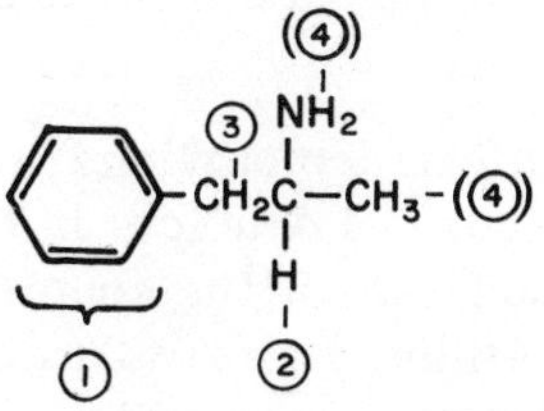

$$C_6H_5\text{-}\overset{③}{CH_2}\text{-}\underset{\underset{②}{H}}{\overset{\overset{(④)}{NH_2}}{C}}\text{-}CH_3\text{-}(④)$$

(phenyl ring: ①)

2. The positions of the signals due to the $-NH_2$ protons are different in the two spectra because the temperatures of the two solutions were different. In the spectrum on p. 158, the $-NH_2$ proton signal accidentally coincides with the low-field member of the methyl doublet. The higher-field position of the $-NH_2$ signal in the spectrum on p. 156 indicates that less hydrogen bonding was taking place when this spectrum was determined. This, in turn, indicates that the temperature of the solution was higher.

   A change in the position of the signal due to the $-NH_2$ protons could also be brought about by a change in the concentration of the solution.

---

## *Answers for Page* 159

1. Band ③ is due to the two nonequivalent methylene protons. The nonequivalence is a consequence of the three different substituents on the adjacent carbon atom. The nonequivalent methylene protons are coupled with each other and with the $\overset{|}{-CH-}$ proton. Although the $\overset{|}{-CH-}$ proton is also coupled with the $-CH_3$ protons, $\Delta\nu_{CH\text{-}CH_3}/J_{CH\text{-}CH_3}$ is about 120/6. Thus, no problems should arise in the $-CH_2-$ band because of "virtual" coupling. Band ③, which can, therefore, be viewed as the AB portion of an ABC or (approximately) ABX system, consists

of two overlapping "AB" patterns. Part of peak a and peaks c, d, and g represent one "AB" pattern, while peaks b, e, f, and h represent the other.

2. The coupling (13 cps) between the nonequivalent methylene protons is equal to the separation of the pairs of doublets in the two "AB" patterns (a–c, d–g, b–e, and f–h). The coupling between the methyl protons and the methine proton (6.1 cps) can also be measured directly. More work is required to extract the coupling constant or coupling constants between the methine proton and the protons on the adjacent carbon atoms. The coupling of the nearly equivalent aromatic protons with each other could not, of course, be determined from this spectrum.

---

*Answers for Page* **161**

1.

(5)
$CH_2CH_2$
(1) { } (1)
C
H $CONH_2$
(4) (2),(3)

2. Band (5) is an $A'_2B'_2$ pattern due to the $-CH_2CH_2-$ protons. The symmetry requires that the arrangement must be $-CH_AH_BCH_AH_B-$ rather than $-CH_AH_ACH_BH_B-$. There are two possible and related causes for this nonequivalence, one of which is slow interconversion between the two different "puckered" forms of the middle ring. However, even with fast interconversion, the nonequivalence could be caused by the different spatial relationships of the ethylene protons with the two substituents in the $\gt C \lt$ (H, $CONH_2$) group.

3. The partial double-bond character of the C–N bond stabilizes the form O$\cdots$C$\cdots$N($H_A$)($H_B$). The proton on the same side as the

carbonyl function ($-H_A$) is shifted to lower field than the proton ($H_B$) on the opposite side. An increase in the temperature of the solution would increase the rate of rotation about the C—N bond and, thus, would cause the difference in chemical shifts between $H_A$ and $H_B$ to become zero.

---

*Answers for Page* 163

1.

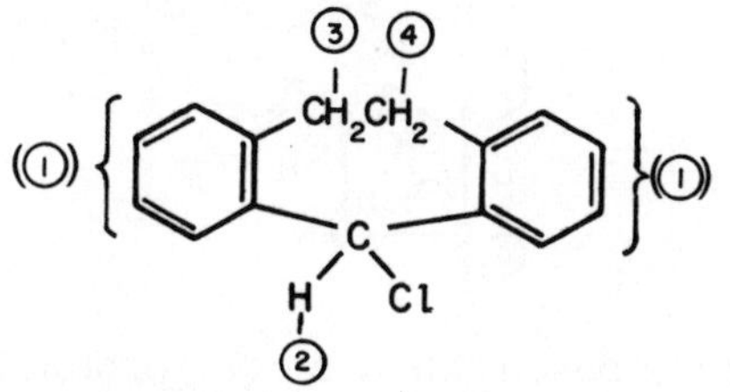

This structure can be deduced by the following reasoning:

| Band | ① | ② | ③ | ④ |
|---|---|---|---|---|
| Number of protons | 8 | 1 | 2 | 2 |

There appear to be three essentially isolated groups of protons: (1) a group of eight aromatic protons; (2) an isolated —CH— proton; and (3) a group of four protons, which gives an $A'_2B'_2$ pattern. The $A'_2B'_2$ pattern does not appear to be due to an ethylene group ($-CH_2CH_2-$) in an aliphatic chain (see scans A and F on p. 112). The compound has C + 1 - ½(H + X) or 15 + 1 - ½(13 + 1) or 9 "rings." There must be at least two phenyl groups present to account for the eight aromatic protons. The two phenyl rings would constitute eight "rings." This suggests that the $-CH_2CH_2-$ and –CH– groups may be incorporated to make a third "ring." The chlorine atom could be substituted on the —CH— group to complete the structure.

2. Bands ③ and ④ constitute an $A'_2B'_2$ pattern. The protons in each of the methylene groups are different from each other

because they differ in their spatial relationship to the –H and –Cl in the –CHCl group.

---

## *Answers for Page* 167

1.

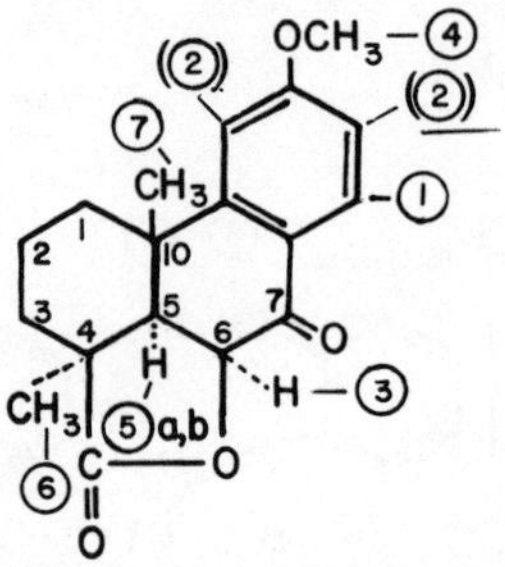

Band ⑤ is superimposed on a single proton signal which is probably due to the equatorial $\alpha$-proton at C-1. This proton would be in the deshielding region (in the plane) of the aromatic ring.

2. The spin systems involved are: bands ① and ②, ABX; bands ③ and ⑤ a and b, AX; bands ④, ⑥, and ⑦ are each due to $A_3$ systems.

3. The factors which must be considered in predicting the vicinal coupling[1] in the AX system are: (1) the dihedral angle, (2) the electronegativities of the substituents, (3) the length of the intervening carbon–carbon bond, and (4) the angles which the bonds to the protons make with the carbon–carbon bond.

---

## *Answers for Page* 169

1. The spectra on pp. 168 and 170 correspond to structures I and II, respectively.

[1] M. Karplus, J. Am. Chem. Soc. 85:2870 (1963).

2.

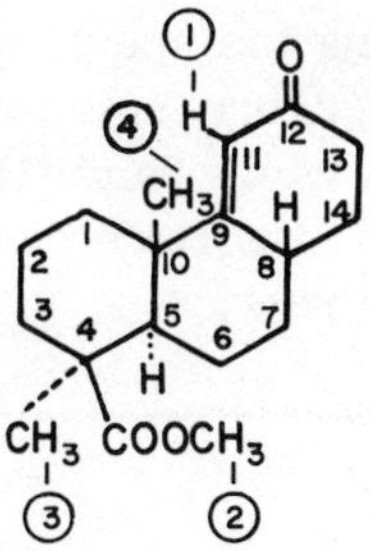

The assignment of bands ③ and ④ is made by noting that the signal at higher field is more sensitive to the change in the location of the double bond.

3. The phasing should have been changed slightly. The curve at the base of each peak (note band ② and the TMS signal) should be at the same level on each side.

---

## *Answers for Page* 171

1.

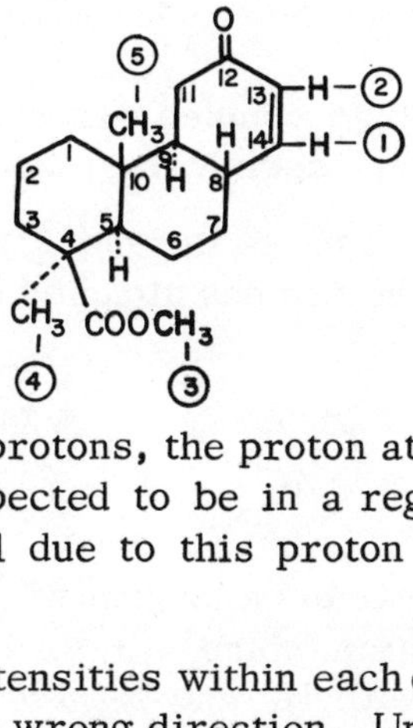

Of the two olefinic protons, the proton at the end of the conjugated system is expected to be in a region of lower electron density. The signal due to this proton would thus occur at a lower field.

2. The relative peak intensities within each closely spaced doublet are distorted in the wrong direction. Under proper operating conditions, the peaks in a multiplet are slanted toward the signal of the proton causing the splitting.

3. The olefinic proton signals were probably saturated, that is, too much RF power was used. Saturation causes the peaks of a multiplet to be enhanced in the order in which they are en-

countered in determining the spectrum. Note that the integrations for bands ① and ② are somewhat low. Band ① corresponds to 0.88 proton, while band ② corresponds to 0.80 proton. The same type of distortion can also be introduced by an excessively high sweep rate.

---

## *Answers for Page* 173

1. The two olefinic protons ($H_A$ and $H_B$) are coupled with each other ($J = 10.3$ cps), as well as to the closest methine proton ($H_X$).

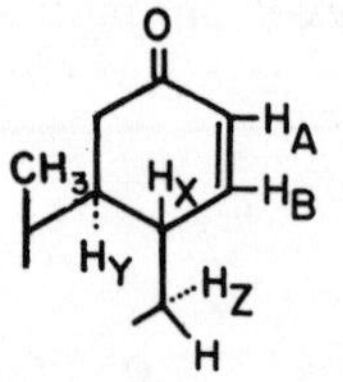

   Note that $H_X$ is in turn coupled strongly with protons $H_Y$ and $H_Z$ which may have chemical shifts very similar to $H_X$.

2. The broadening of the peaks in band ② may be due to coupling with one or both of the protons attached to C-11:

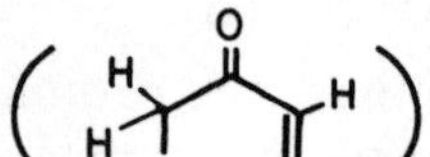

   Such couplings appear to be largest when the protons involved are both in the same plane. If the broadening were due to "virtual" coupling with $H_Y$ and $H_Z$, the signals of both $H_A$ and $H_B$ would be expected to show essentially the same broadening, because each of these protons is coupled with $H_X$ to the same extent.

3. The magnitude of an allylic proton–proton coupling ($CH_X{=}C{-}CH_A$) depends primarily on the angle that the bond between the saturated carbon atom and $H_A$ makes with the

plane of the double bond. In this example, the coupling is relatively large because the angle involved is about 90°.

---

*Answers for Page* 175

1.

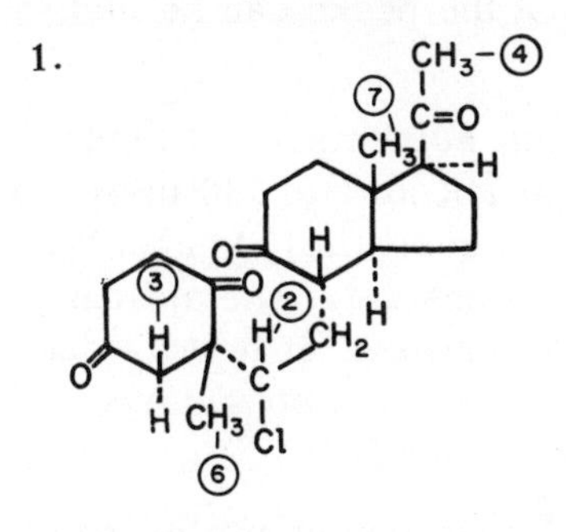


2. The absence of band (3) in the acetate II indicates that this one-proton band is probably due to a proton on or near ring A. The only proton that could give a clean doublet would be one of the protons in the isolated methylene group. Since there are three different substituents on the adjacent atom, these two protons could give an AB or AX pattern.

3. Yes, it is reasonable to expect some enhancement of the geminal $\left( >\mathrm{C}<^{\mathrm{H}}_{\mathrm{H}} \right)$ coupling because of the adjacent carbonyl.[1] The "normal" coupling constant of −12.5 cps has been enhanced by −3 cps. This is consistent with the predicted increment for an angle of either approximately 8° or 52° between the methylene group and the $\pi$-bond.

4. The separation of the outer peaks should be approximately equal to $2(J_{AB})/(1$ − the ratio of the intensity of the smaller to that of the larger peak). This value is equal to $2(15.5)/(1 - 0.60)$ or 78 cps.

---

[1] M. Barfield and D. M. Grant, J. Am. Chem. Soc. 85:1899 (1963).

## *Answers for Page* 177

1. No, the separations in band (2) cannot be assumed to be coupling constants. This may be part of a higher-order pattern which would have to be analyzed completely in order to extract the coupling constants. This analysis would be impossible from this single spectrum because not all of the peaks can be identified.

2. The spectrum could be run in various solvents or at higher frequencies using, for example, a Varian HR- or HA-100 instrument. The small splitting would remain essentially unchanged if it did, indeed, represent a coupling constant. The splitting might change to a measurable degree, however, if it were due to "virtual" coupling. This type of problem is considered in detail in conjunction with a question on p. 199.

3. The proton which gives rise to band (2) is coupled to at least one proton. It is not coupled equally to the two protons in the adjacent methylene group. If it were equally coupled to these two protons, a triplet would result. The smaller splittings in band (2) may be due to either real or "virtual" coupling.

---

## *Answers for Page* 179

1. No, the sample is not pure. The extra peaks on the low-field side of angular methyl signals (3) and (4) indicate the presence of a closely related compound. This impurity is

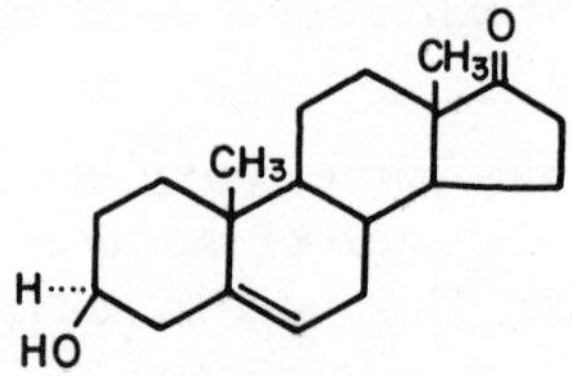

$C_{19}H_{28}O_2$

The maximum long-range coupling of the angular methyl signals which should be anticipated is about 1 cps.[1]

[1] See pp. 116-121 of the text by Bhacca and Williams.

2. The peak to the left of the chloroform signal is probably due to benzene.

3.

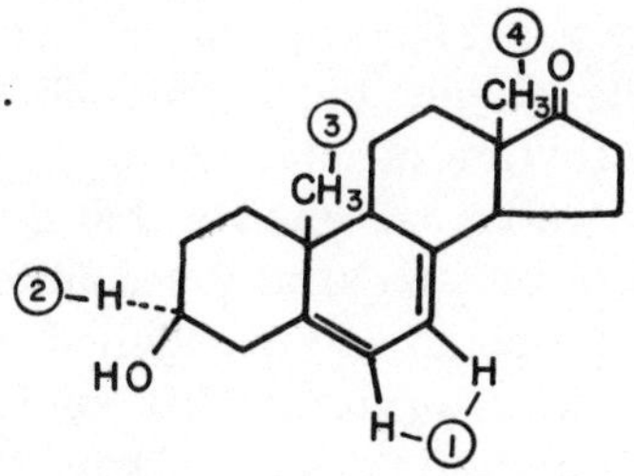

4. The two olefinic protons have very similar chemical shifts. The single peak in band (1) is the combination of the two major peaks of an AB pattern. The outer peaks have been lost in the noise because of the small value of the ratio $\Delta\nu_{AB}/J_{AB}$.

---

## *Answers for Page* 181

1.

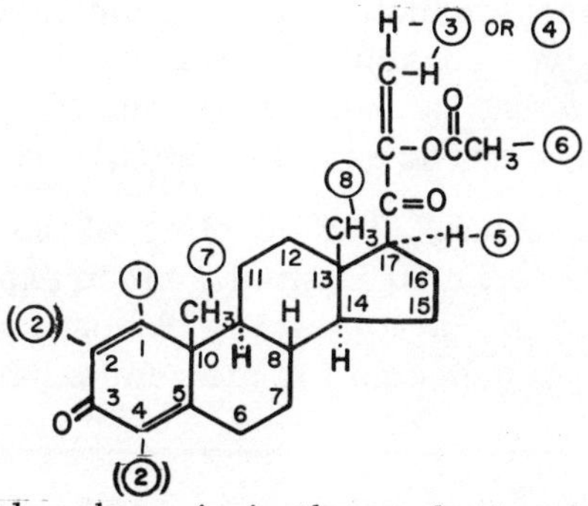

The chief problem here is in the analysis of the patterns seen in the olefinic region. First, it is noticed that the small peak on the low-field (left) side of band (1) is due to $CHCl_3$. The vicinal olefinic protons of a cyclohexene system, such as the protons at C-1 and C-2, are usually coupled to the extent of approximately 10 cps. The coupling between the protons at C-1 and C-4 would be expected to be very small. The C-1 proton should thus appear as a clear doublet having a separation of 10 cps. Therefore, it is reasonable to assign band (1) to the proton at C-1.

Band (2) can easily accommodate the signals expected from the proton at C-2. The pattern consists of a doublet corresponding to the doublet in band ①, each member of which is split further by a small coupling to the proton at C-4.

Bands ③ and ④ can be reasonably assigned to an AB system because the splittings are the same, and each doublet shows the expected higher-order distortion of intensities.

The only remaining proton signal to be assigned is that due to the C-4 proton. This is expected to be essentially a doublet due to the small coupling to the C-2 proton and broadened by coupling primarily with the C-6 β (axial) proton. This signal due to the C-4 proton must be in the high-field side of band ② along with the high-field portion of the signal due to the proton at C-2.

In actual practice, the pattern seen in bands ① and ② is easily recognized as being typical of the three protons in a 3-oxo-$\Delta^{1,4}$ system. This leaves only bands ③ and ④ to be interpreted.

The triplet in band ⑤ is typical of a C-17 proton pattern in many steroids in which this proton is coupled with only the C-16 methylene protons. The approximate positions of the C-18 and C-19 angular methyl protons can be calculated using Zürcher's additive constants. Since this particular side chain is not listed, a substitute such as β-$COOCH_3$ or β-$COCH_2OCOCH_3$ must be used for the approximation.

2. Structure II is correct. The coupling of the olefinic protons in the side chain (2.2 cps) is smaller than expected for either a cis (5–12 cps) or a trans (12–18 cps) coupling, but is in the range (0–3 cps) expected for the geminal coupling.

---

*Answers for Page* 183

1.

2. Bands ② and ③ constitute the eight expected peaks in the AB portion of an ABX pattern $\left(\begin{array}{c} CH_AH_BF_X \\ | \\ C{=}O \\ | \end{array}\right)$. The two methylene protons are coupled with each other and with the fluorine atom (the spin quantum number of fluorine is ½). The non-equivalence of the two methylene protons may be due to the presence of three different substituents on the nearby carbon atom (C-17) or to restricted rotation.

Note that the weaker members of the two "AB" patterns in the spectrum on p. 182 are lost in the noise. Bands ② and ③ could be confused with a weakly coupled, simple AB pattern. The geminal "coupling constant" would be mistakenly assumed to be 5 cps, whereas the actual coupling constant is 16.5 cps. The slight apparent slanting of the "doublets" toward each other in the trace on p. 182 is due to noise and is accidental. The peaks in the doublets should be of the same height. Note the ratios on p. 184. The apparent slanting seen on p. 182 could add to the possible confusion with a simple AB pattern.

The sharp peak in band ④ is due to the C-16 proton, which is coupled only slightly with the two neighboring protons.

---

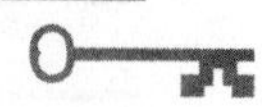

## *Answers for Page* 185

1. The magnitude of the coupling ($J_{AB}$) between the two methylene protons (16.5 cps, assumed to be negative) is equal to the separations a–b and c–d in each of the two bands. The separations of the first and third peaks (a–c and b–d) in each of the two "AB" patterns are equal. The variables are related as indicated by the following diagram:

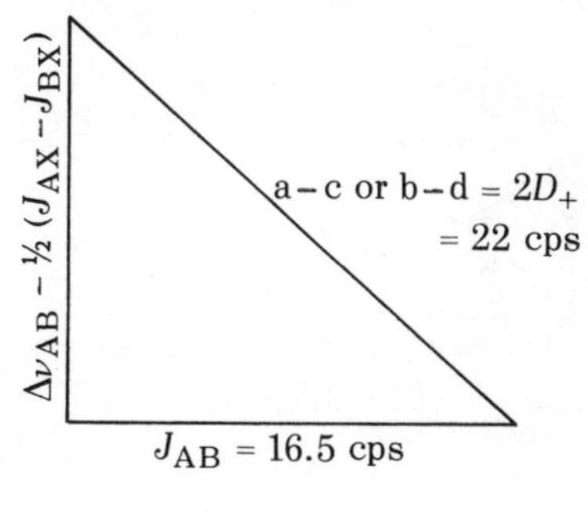


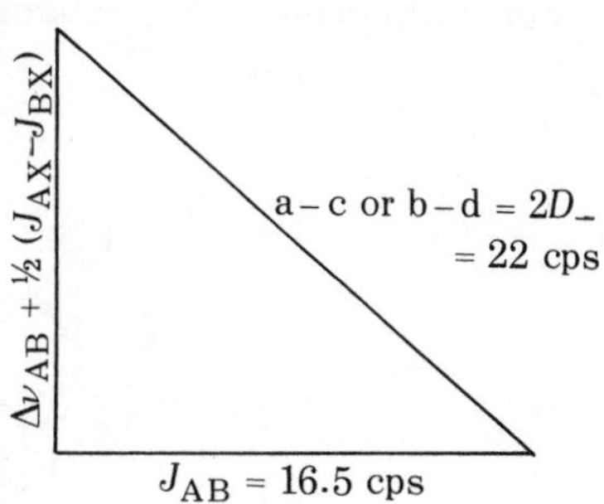

It is clear that the two protons are nonequivalent because otherwise the signal due to these two protons could only be a doublet. The equality of the two vertical sides of the two triangles means that $-\frac{1}{2}(J_{AX} - J_{BX})$ must be equal to $+\frac{1}{2}(J_{AX} - J_{BX})$, which, in turn, means that $J_{AX} = J_{BX}$. Of course, this is to be expected in the grouping $-CH_AH_BF$. The calculated vertical sides of the two triangles are thus both equal to $\Delta\nu_{AB}$. The separation (47.8 cps) of the centers of the two "AB" patterns (318.3 and 270.5 cps) must be equal to $\frac{1}{2}(J_{AX} + J_{BX})$ or, since $J_{AX} = J_{BX}$, to either $J_{AX}$ or $J_{BX}$.

2. The fluorine spectrum would consist of a 1:2:1 triplet having a spacing of 47.8 cps. The middle peak would be composed of the two superimposed peaks (separated by $2D_+ - 2D_-$), while the outer peaks would be separated by $J_{AX} + J_{BX}$. The other two possible peaks would be separated by 44 cps ($2D_+ + 2D_-$), but they would have zero intensities ($2\phi_+ = 2\phi_-$, so that $\sin^2[\frac{1}{2}(2\phi_+ - 2\phi_-)] = 0$).

   It should be noted that in this analysis only $J_{AX} + J_{BX}$ could have been extracted from the X pattern alone. Frequently, it is necessary to examine the X pattern in order to cross-check the values of $2D_+ - 2D_-$, $2D_+ + 2D_-$, and $J_{AX} + J_{BX}$ which are obtained from the AB portion.

---

## *Answers for Page* 187

1. The sharp peak ⑤, which shows good ringing, is probably due to methanol. This would be confirmed if the size of peak ⑤ were increased in a spectrum repeated after the addition of a small amount of methanol. Most of the methanol was washed out during the $D_2O$ exchange.

2.

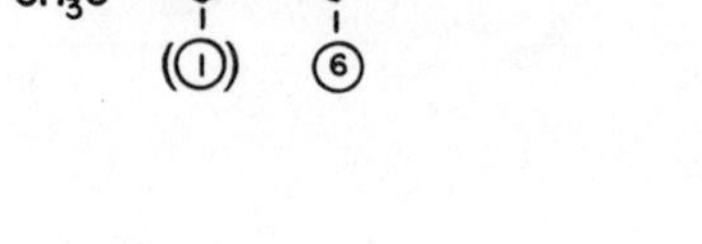

The signals near 430 cps (part of band (1)) are due to the C-1 proton which is coupled to the protons on C-2 and C-4.

3. Not all of the stereochemical problems in the proposed structure are solved by this spectrum. The olefinic proton which causes band (2) is coupled not only with the $-CH_2Br$ protons, but also with the protons on C-16. The allylic couplings would be very similar, however (within 0.5 cps), for both the cis and the trans isomers. If both isomers were available, then perhaps an assignment could be made.

   Although there may be a small difference between the chemical shifts of the methylene protons in the $\underset{\|}{C}HCH_2Br$ group, these protons are strongly coupled ($\sim -12$ cps) and, thus, behave as if they are equivalent. On p. 112 of their text, Bhacca and Williams show how the difference in chemical shifts in a similar grouping ($\underset{\|}{C}HCH_2OH$) becomes large enough at 100 Mcps to cause the system to behave as an ABX type.

---

## *Answers for Page* 189

1. The spectrum on p. 188 belongs to the pure isomer. Note the doubling of the signals due to olefinic (band (1)), methoxy (band (3)), and methyl (band (4)) protons in the spectrum on p. 190. The assignments of the bands in the two spectra are as follows:

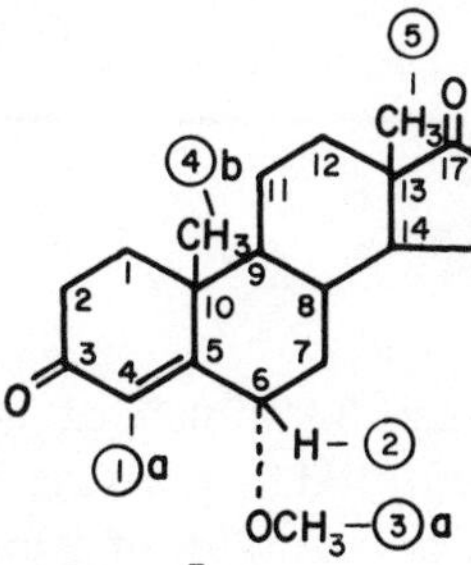

I 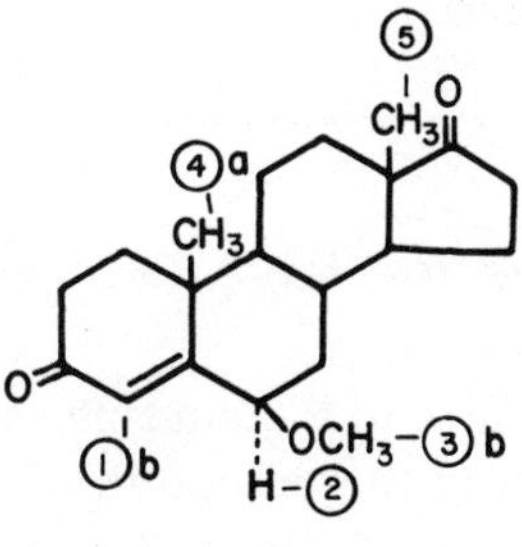          II

2. Isomer I was obtained pure. There are three distinct ways in which this can be established.

First, the olefinic proton signal (band ①a) in the pure isomer is a doublet, and, in the other, this signal (①b) is a singlet. Maximum allylic coupling (1–3 cps) is observed when the bond to the allylic proton is perpendicular to the plane of the double bond. Minimum coupling is observed when the allylic proton is in the same plane as the double bond. The pure isomer must, therefore, have a C-6 axial (essentially perpendicular to the plane of the A and B rings) proton which corresponds to isomer I.

Second, the relative chemical shifts of the olefinic, methoxy, and methyl protons can be used to help choose either I or II. In isomer I, the signal due to the olefinic proton should be at lower field than the corresponding signal in II because of the dipole moment of the methoxy group. The signal due to the methoxy protons should also be to the left because the methoxy group is in the deshielding zone of the C-4, 5 double bond. The 1,3-diaxial relationship of the C-6 methoxy and the C-19 methyl group in isomer II should, because of the dipole moment of the methoxy group, cause the signal due to the C-19 methyl group in II to be further downfield than the C-19 methyl signal in I.

Third, in isomer I, the C-6 axial proton is coupled more strongly with the protons on C-7 than is the equatorial proton C-6 in isomer II. This makes band ② for isomer I broader than band ② for isomer II. The almost equal dihedral angle between the C-6 proton and each of the two C-7 protons for isomer II causes band ② to be essentially a triplet. An expansion of band ② for isomer I is shown and discussed on pp. 194 and 195.

---

*Answer for Page* **191**

1. The mixture contains approximately 33 mol.% of isomer I. One way the composition can be calculated is by using the integrations of bands ③a and ③b. The percentage of I is equal to (100) (7)/(7 + 14.5) or 33 mol.%. The heights of bands ③a and ③b could also be used for this estimation.

---

## *Answer for Page* 193

1. Two differences caused by the reverse sweeping are as follows: (1) The ringing occurs on the low-field side of the strong peaks rather than on the high-field side. (2) The relative intensities of the doublet in band ① are reversed. This reversal of intensities indicates that the power level was too high for this particular proton (saturation was occurring). Fast sweeping, even with a low power level, can also cause this same type of distortion.

---

## *Answers for Page* 195

1. The system involved is best approximated as an ABCMX type.

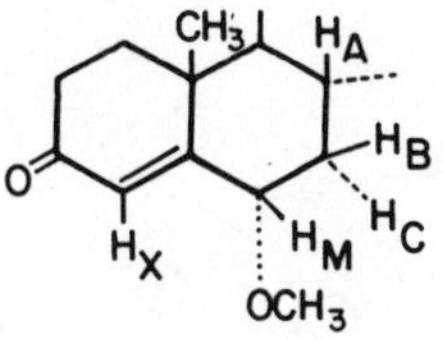

2. Only the smallest splitting (2.0 cps, a–b, c–d, e–f, and g–h) due to the coupling between $H_M$ and $H_X$ can be taken as a coupling constant. This smallest splitting is clearly due to a first-order coupling. The centers of the doublets a–b and g–h are separated by $J_{AM} + J_{MB} + J_{MC}$. The separation between the centers of the doublets a–b and c–d (and e–f and g–h) should not be taken as coupling constants, because the ratio of the differences in the chemical shifts between $H_A$, $H_B$, and $H_C$ and the various coupling constants between these nuclei cannot be determined from this spectrum.

---

*Answers for Page* 197

1.

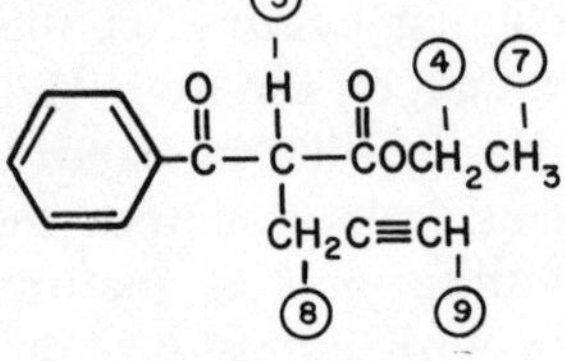

$C_6H_5$–C(=O)–CH(③)(–$CH_2$(⑧)C≡CH(⑨))–C(=O)O$CH_2$(④)$CH_3$(⑦)

Band ⑤ is due to water which was present in the pyridine.

2.

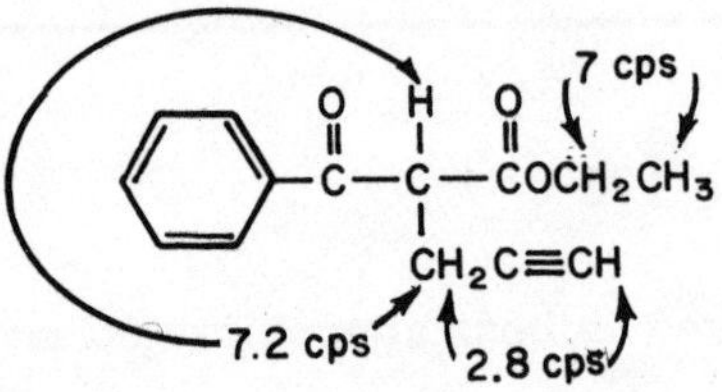

$C_6H_5$–C(=O)–CH($CH_2$C≡CH)–C(=O)O$CH_2CH_3$

7 cps (between $CH_2$ and $CH_3$ of the ethyl group); 7.2 cps (between H and $CH_2$); 2.8 cps (between $CH_2$ and ≡CH)

---

*Answers for Page* 199

1.

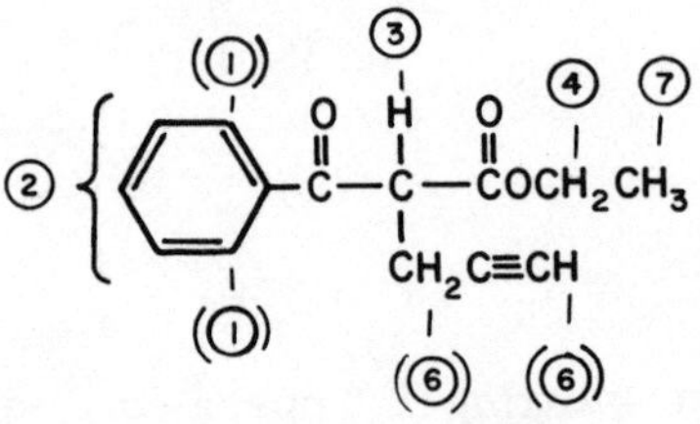

②{$C_6H_5$ (①, ①)}–C(=O)–CH(③)(–$CH_2$(⑥)C≡CH(⑥))–C(=O)O$CH_2$(④)$CH_3$(⑦)

Band ⑤ is due to water which was present in the solvent. The multiplet on the high-field (right) side of band ⑥ is due to $CHD_2S(=O)CD_3$. The small band on the low-field (left) side of band ⑥ is due to an impurity.

2. The following four factors should be considered when a band consists of more peaks than would normally be predicted: (1) The assignment of the bands may be incorrect; (2) closely

related impurities may be present; (3) there may be unexpected nonequivalence of protons; and (4) the extra peaks may be due to higher-order effects ("virtual" coupling). In this example, the assignment of band ③ to the methine (–CH–) proton is firmly based on the integration and position of the band. The only other single-proton band should be due to the acetylenic proton. The signal due to this acetylenic proton is very reasonably assumed to be in band ⑥. Band ⑥ also includes the signals due to the protons in the methylene attached to the acetylenic group. Proof that closely related impurities are absent is often difficult to obtain. Other data obtained on this sample indicate that the compound was reasonably pure. The spectrum of the same sample in pyridine (p. 196) shows only the multiplicities expected by first-order rules. Two different aspects of unexpected nonequivalence should be considered here. First, there are three different substituents on the carbon adjacent to the methylene group in the $-CH_2C{\equiv}CH$ portion. This could conceivably lead to nonequivalence of these methylene protons. This, in turn, would mean that the methine (–CH–) proton was the X of an unisolated ABX system. The X pattern would still consist of a triplet, however, if the coupling between the methine proton and the two methylene protons were equal. This would be the case if there were fast rotation and equal populations in the conformations about the bond between the methine and methylene ($-CHCH_2-$) groups. This possibility of slow rotation about the bond between the methine and methylene groups is the second aspect of unexpected nonequivalence which must be considered. If each of the three stable conformations were equally populated and there were slow rotation about the bond between the –CH– and $-CH_2-$ groups, there could be three different and equally intense patterns. These could overlap and produce considerable complexity. A small value for the ratio $\Delta\nu/J$ in the $-CH_2C{\equiv}CH$ portion could cause extra peaks (due to "virtual" coupling) in band ③. The $-CHCH_2-$ portion then could not be considered to be first-order, because of the further coupling of the $-CH_2-$ and $-C{\equiv}CH$ protons. Clearly, the value of the ratio $\Delta\nu/J$ in the $-CH_2C{\equiv}CH$ portion is not large in the spectrum on p. 198, but the complexity of this band (⑥) makes the actual values difficult to extract. In the spectrum which was determined using pyridine (p. 196), the ratio $\Delta\nu/J$ in the $-CH_2C{\equiv}CH$ group is 24/2.6. Band (3) in the pyridine spectrum is a clean triplet. This does not prove, however, that the extra peaks in the spectrum in deuterodimethylsulfoxide are due to "virtual" coupling (see the answer to question 3 below).

3. Any experiment which would increase the ratio $\Delta\nu/J$ in the $-CH_2C{\equiv}CH$ group or increase the rate of rotation about the bond between the $-\overset{}{\underset{|}{C}}H-$ and $-CH_2-$ groups would help to explain the multiplicity of band (3). There would be pitfalls in the interpretation of the results of many of these experiments. The best experiment to perform is the synthesis of the compound in which the acetylenic proton is replaced by deuterium. The acetylenic proton signals in band (6) would then be absent and, at the same time, the ratio $\Delta\nu/J$ in the $-CH_2C{\equiv}CD$ group would be extremely large. If the complexity of band (3) in the spectrum on p. 198 were due to "virtual" coupling, the replacement of the acetylenic proton by deuterium would remove this complexity. If the complexity were due to slow rotation, however, the replacement by deuterium would not appreciably affect the appearance of band (3). The ratio $\Delta\nu/J$ in the $-CH_2C{\equiv}CH$ group would be increased by $^{100}/_{60}$ by using a 100-Mcps instrument. This method would not change the rate of rotation about the bonds. The ratio $\Delta\nu/J$ still might not be large enough, however, to remove all of the "virtual" coupling. Increasing the ratio $\Delta\nu/J$ in the $-CH_2C{\equiv}CH$ group by changing the solvent (p. 196) does, indeed, cause simplification of band (6). The simplification could also be due, however, to an increased rate of rotation about the bonds or to a change in the populations in the various conformations. Elevation of the temperature should cause an increase in the rate of rotation about the bond between the carbon atoms in the $-\underset{|}{C}HCH_2-$ groups. This, in turn, would lead to simplification of band (3) if the complexity were due to slow rotation. The interpretation of the results of this experiment could be misleading if changes in chemical shifts also occurred. The position of the acetylenic proton signal would be particularly sensitive to any change in hydrogen bonding. If the complexity of band (3) were due to "virtual" coupling, a double-resonance experiment, in which band (3) was irradiated while band (6) was being observed, should

cause band ⑥ to collapse to a single $AB_2$ pattern. If the complexity of band ③ were due to slow rotation, band ⑥ would probably, though not necessarily, collapse to give two or three different and overlapping $A_2B$ patterns. This double-resonance experiment should be performed using a frequency sweep rather than a field sweep. This would require an instrument other than the Varian A-60.

---

## *Answers for Page* 201

1. This molecule is so large that free tumbling is prevented. Free tumbling is necessary in order to average out the effects of local magnetic fields which result from neighboring protons. (This effect is called dipole–dipole broadening.)

2. [Structure: $C_6H_5$–$CH_2O$–C(=O)ND–CH(CH(CH_3)_2)–C(=O)ND–CH($CH_2$–$C_6H_4$–OD)–C(=O)ND–CH(CH($CH_3$)$CH_2CH_3$)–C(=O)ND–CH($CH_2$–imidazolyl(ND))–C(=O)–$OCH_3$, with band labels ① to ⑩]

Bands which include two or more signals are placed in parentheses. The signals due to all of the –NH– protons have been removed by exchange with the $CD_3COOD$. Band ⑨ is due to $CHD_2COOD$. Note the distortion in this quintet which is due to the use of too much radio-frequency power for this band.

These assignments can be made most easily in the following order: First, the signals due to $CHD_2$–, [phenyl], $-CH_2O-$, and $-OCH_3$ are identified. By noting the integration of the band due to the $-OCH_3$ protons (⑦), it can be seen that band ① corresponds to one proton. This must be the [imidazole C-2 H] proton. The [imidazole C-4 H] proton must accidentally coincide with band ②. This is confirmed by the integration. The $A'_2B'_2$ pattern in band ③ can then be assigned to the *p*-disubstituted phenyl protons. The protons in the $-CH(CH_3)_2$ and $-\underset{\underset{\textstyle CH_2CH_3}{|}}{CH}-CH_3$ groups must be in band ⑩. This leaves only bands ⑤, ⑥, and ⑧ to be assigned. Band ⑥ is assigned to the only two protons that would give rise to doublets. Band ⑥ might be confused with the X pattern of an ABX system if it were not for the integration curve which shows that, relatively, the signal is due to two protons. The low-field doublet is also split slightly more than the high-field doublet. These splittings would be equal in the X pattern of an ABX system. The protons next to the aromatic systems

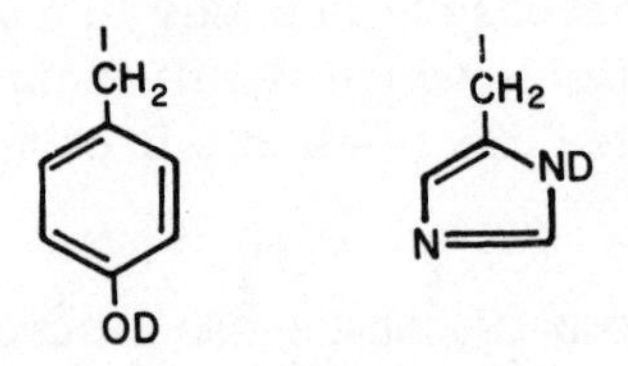


may both be AB portions of ABX patterns. The two broad bands in ⑧ are in the expected positions for these protons. This now leaves band ⑤ to be assigned to the protons which correspond to the X portions of these two ABX patterns.

---

# APPENDIX A

Characteristic positions of various proton signals at 60 Mcps with reference to internal TMS. Unless otherwise noted, the positions are those of aliphatic protons. The data are from the text by L. M. Jackman and the Varian catalogs by Bhacca, Johnson, and Shoolery and by Bhacca, Hollis, Johnson, and Pier.

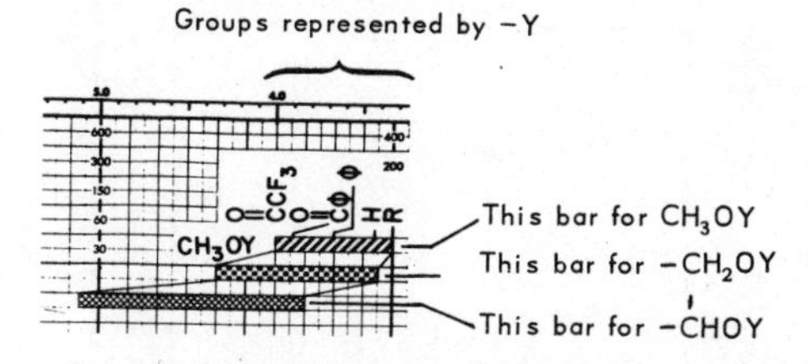


All of the protons in this group are on a carbon atom attached to an oxygen atom. The groups represented by —Y are shown above the top bar. In general, the substituents represented by —Y (or —X in the case of the halogens) cause the same relative shifts of $CH_3$—, —$CH_2$—, and —CH— protons.

In some cases, in HC≡CY for example, only one type of proton is represented. The shading of the bars indicates whether $CH_3$—, —$CH_2$—, or —CH— protons are involved.

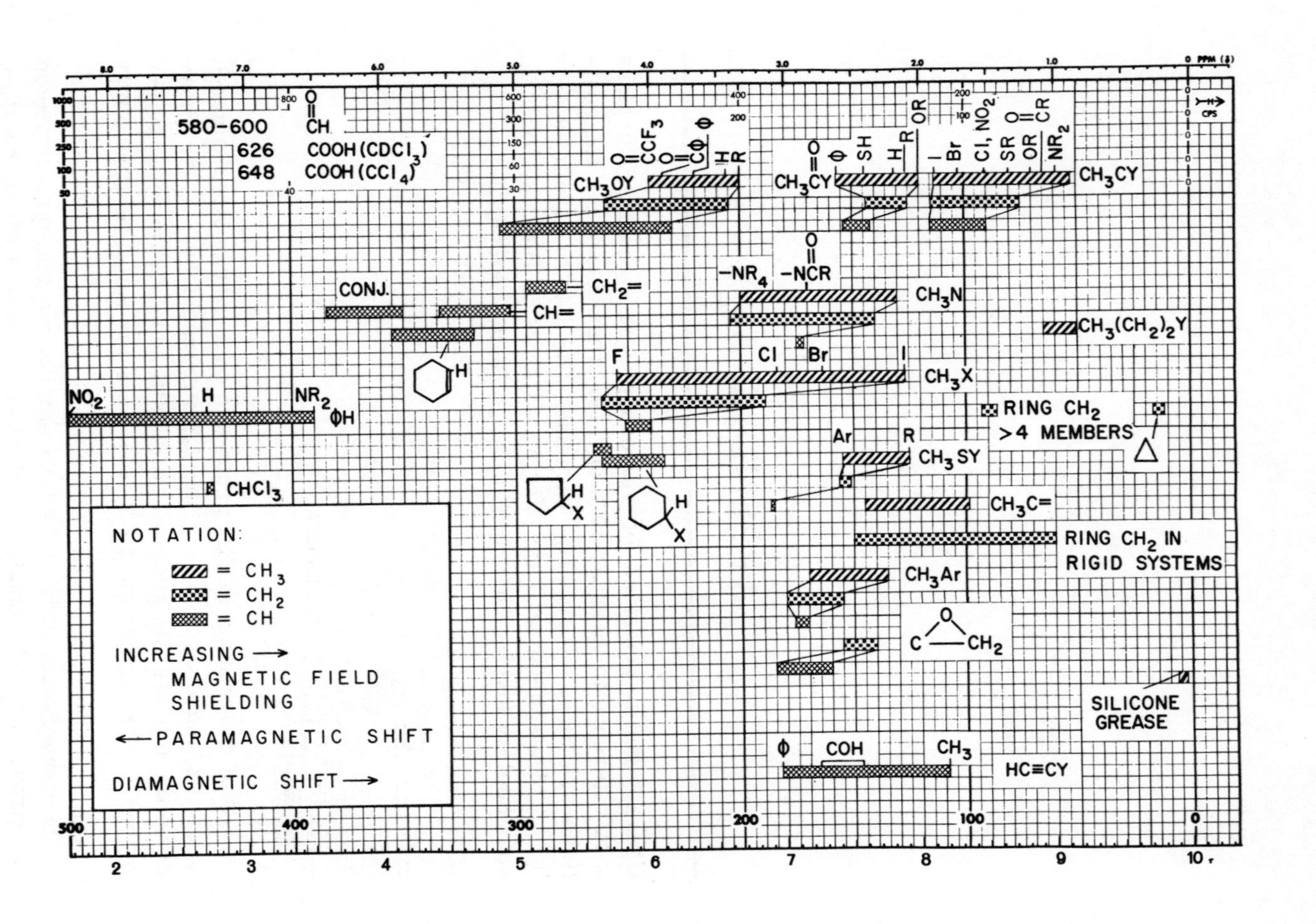

NOTATION:
= CH3
= CH2
= CH
INCREASING → MAGNETIC FIELD SHIELDING
← PARAMAGNETIC SHIFT
DIAMAGNETIC SHIFT →
580-600
626
648
COOH (CDCl3)
COOH (CCl4)
NO2
H
NR2
ΦH
CHCl3
CONJ.
CH=
CH2=
CH3OY
CH3CY
CH3N
CH3X
F
Cl
Br
I
-NR4
-NCR
CH3SY
Ar
R
CH3C=
RING CH2 IN RIGID SYSTEMS
RING CH2 >4 MEMBERS
CH3(CH2)2Y
CH3Ar
CH3
COH
HC≡CY
Φ
SILICONE GREASE

# APPENDIX B: Shoolery's Additive Constants

The position in cps of the signal due to an aliphatic methylene group in $XCH_2Y$ in $CCl_4$ is equal to the sum of the constants (cps) for the two substituents plus 14 cps. The position in ppm is equal to the constants (ppm) plus 0.233 ppm. The constants may be used, but with much less success, for methine-proton positions. These data, which are for 60 Mcps with TMS as an internal reference, are taken from Varian Associates Tech. Information Bulletin, Vol. 2, No. 3; L. M. Jackman, Applications of Nuclear Magnetic Resonance Spectroscopy in Organic Chemistry, Pergamon Press (New York), 1959, p. 59; and R. M. Silverstein and G. Clayton Bassler, Spectrometric Identification of Organic Compounds, John Wiley & Sons (New York), 1963, Appendix B.

| Group | Additive constants (cps) | Additive constants (ppm) |
|---|---|---|
| $-CH_3$ | 28 | 0.47 |
| $-CF_3$ | 68 | 1.14 |
| $>C=C<$ | 79 | 1.32 |
| $-C\equiv C-$ | 86 | 1.44 |
| $-\overset{O}{\overset{\parallel}{C}}OR$ | 93 | 1.55 |
| $-NR_2$ | 94 | 1.57 |
| $-\overset{O}{\overset{\parallel}{C}}NR_2$ | 95 | 1.59 |
| $-SR$ | 98 | 1.64 |
| $-C\equiv N$ | 102 | 1.70 |
| $-\overset{O}{\overset{\parallel}{C}}R$ | 102 | 1.70 |
| $-I$ | 109 | 1.82 |
| $-\phi$ | 110 | 1.83 |
| $-Br$ | 140 | 2.33 |
| $-O$ alkyl | 142 | 2.36 |
| $-Cl$ | 152 | 2.53 |
| $-OH$ | 153 | 2.56 |
| $-O\overset{O}{\overset{\parallel}{C}}R$ | 188 | 3.13 |
| $-O\phi$ | 194 | 3.23 |

# APPENDIX C: Major Factors Which Affect the Magnitude of Coupling Constants

Note that magnitude only is considered. Geminal $\left(\begin{matrix}H\\ \\H\end{matrix}\!\!>\!C\right)$ coupling constants are usually negative so that an increase in magnitude is brought about by a negative contribution. M. Barfield and D. M. Grant present a summary of the "Theory of Nuclear Spin–Spin Coupling" on pp. 149-193 in Advances in Magnetic Resonance, Vol. 1 (J. S. Waugh, ed.), Academic Press (New York), 1965.

[a] For leading references and a general theory, see J. A. Pople and A. A. Bothner-By, J. Chem. Phys. 42:1339 (1965). [b] M. Barfield and D. M. Grant, J. Am. Chem. Soc. 85:1899 (1963). [c] M. Karplus, J. Am. Chem. Soc. 85:2870 (1963). [d] For a general discussion of long-range spin–spin coupling, see S. Sternhell, Rev. Pure Appl. Chem. 14:15 (1964). [e] K. B. Wiberg, B. R. Lowry, and B. J. Nist, J. Am. Chem. Soc. 84:1594 (1962). [f] A. Rassat, C. W. Jefford, J. M. Lehn, and B. Waegell, Tetrahedron Letters, 233 (1964). [g] N. S. Bhacca and D. H. Williams, Applications of NMR Spectroscopy in Organic Chemistry (Illustrations from the Steroid Field), Holden-Day, Inc. (San Francisco), 1964, p. 121.

*Magnitude of* $J_{AB}$

| $H_A$ ($\alpha$) $>$C [a] $H_B$ |
|---|

Increases with (1) decreasing s-character of carbon atom, (2) decreasing electronegativity of substituent on carbon atom, (3) increasing electronegativity of the substituent on the atom attached to the carbon atom, and (4) presence of adjacent pairs of $\pi$-electrons (largest when the line between the protons is perpendicular to the nodal plane of the $\pi$-electron system) [a,b].

When substituted with —O< or —N< , $J_{AB}$ is a function of stereochemistry (maximum decrease when the H–H axis is perpendicular to the C—O—R or C—N—R plane).

Very little correlation with angle $\alpha$ except in unsubstituted hydrocarbons where $J_{AB}$ increases with an increase of the angle.

| $CH_A-CH_B$ [c] |
|---|

Function of dihedral angle (angle between the planes defined by $CH_A$—C and C—$CH_B$) (maximum for 0° and 180°; minimum for 90°).

Decreases with increasing $H_A$–C–C and C–C–$H_B$ angles, C—C bond length, and increasing electronegativity of substituents.

| $CH_A{=}CH_B$ |
|---|

Larger for trans than for cis. Decreases with increasing $H_A$–C=C and C=C–$H_B$ angles, increasing electronegativity of substituents, and decreasing $\pi$-bond order.

| $CH_ACCH_B$ [d,e,f] |
|---|

Sensitive to stereochemistry. Often maximum when in the planar "W" shape ($H_A$–C–C–C–$H_B$).

$$\mathrm{CH_A\overset{\overset{\displaystyle O}{\|}}{C}CH_B}$$ [d,g]

Maximum when the system is planar.

$H_A{-}C{-}C{=}C{-}H_B$ [d]

Maximum when the angle between the planes defined by $H_A{-}C{-}C{=}$ and $-C{=}C{-}H_B$ is 90°; minimum when this angle is 0°. Approximately the same for cis and trans.

$H_A{-}C{-}C{=}C{-}C{-}H_B$ [d]

Maximum when the planes defined by $H_A{-}C{-}C{=}$ and $=C{-}C{-}H_B$ both make an angle of 90° with the plane defined by $C{-}C{=}C{-}C$. Approximately the same for cis and trans.

# APPENDICES D, E, AND F

Most of the values in the three charts that follow are taken from the collections given in the texts by Jackman and by Pople, Schneider, and Bernstein and in a private communication from D. P. Hollis. An extensive compilation of coupling constants is given and discussed by A. A. Bothner-By in the chapter, "Geminal and Vicinal Proton–Proton Coupling Constants in Organic Chemistry," in Advances in Magnetic Resonance, Vol. 1 (J. S. Waugh, ed.), Academic Press (New York), 1965, pp. 195-316. In addition to the following, see the references given in Appendix C.

A survey of couplings between vicinal protons in cyclohexane systems is given by A. C. Huitric, J. B. Carr, W. F. Trager, and B. J. Nist, Tetrahedron 19:2145 (1963). The effect of a carbonyl on the coupling between geminal protons is discussed by T. Takahaski, Tetrahedron Letters, No. 11, 565 (1964), and the effect of ring size on coupling in the 1,3-dioxolans is discussed by T. A. Crabb and R. C. Cookson, Tetrahedron Letters, No. 12, 679 (1964). Many leading references are given by E. O. Bishop, Ann. Repts. 58:55 (1961).

A review of the NMR of heterocyclic compounds is given by R. F. M. White in Chapter 9 of Physical Methods in Heterocyclic Chemistry (A. R. Katritzky, ed.), Vol. II, Academic Press (New York), 1963.

*Appendix D*

Spin–spin coupling constants for protons on saturated systems. The most commonly observed ranges are indicated by boxes. See p. 281 for references and Appendix C for the major factors which affect the magnitudes of these coupling constants.

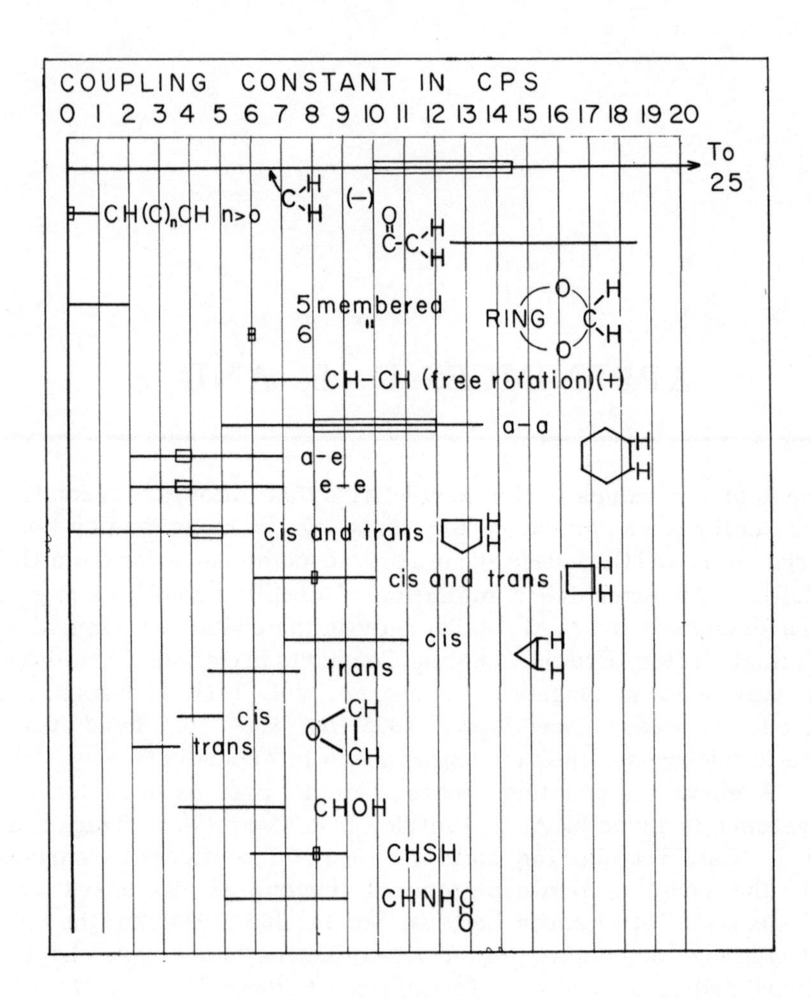

*Appendix E*

Spin–spin coupling constants for aldehydic protons and protons on multiple bonds. The most commonly observed ranges are indicated by boxes. See p. 281 for references and Appendix C for the major factors which affect the magnitudes of these coupling constants.

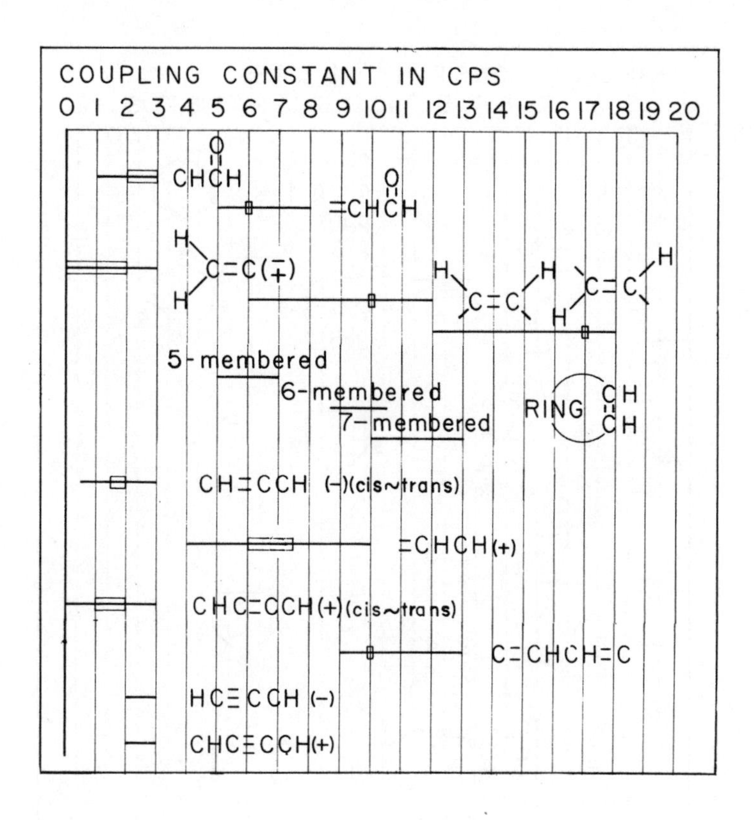

*Appendix F*

Spin–spin coupling constants for protons on aromatic systems. The most commonly observed ranges are indicated by boxes. See p. 281 for references and Appendix C for the major factors which affect the magnitudes of coupling constants.

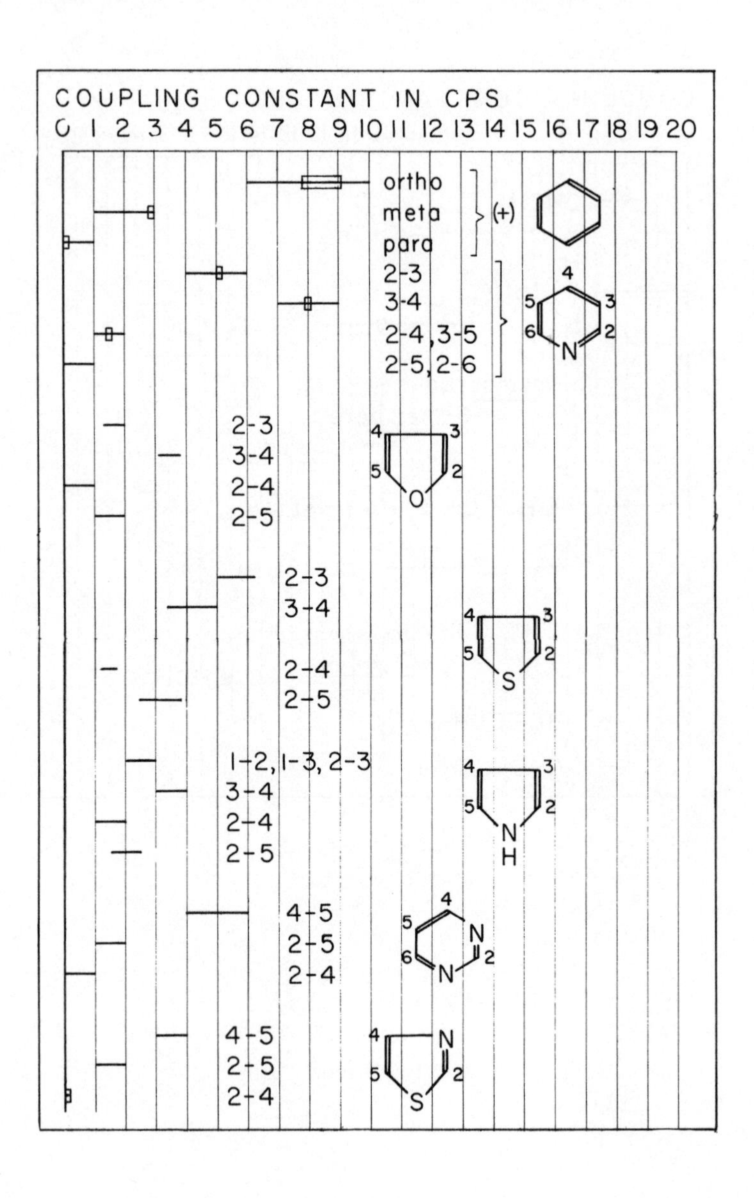

# MOLECULAR FORMULA INDEX

Spectra of compounds after $D_2O$ exchange are listed under the parent structures.

| Molecular formula | Structure | Page number |
|---|---|---|
| $C_2H_6OS$ | $CH_3\overset{O}{\overset{\|}{S}}CH_3$ | 72 |
| $C_2H_7O_3P$ | $H\overset{O}{\overset{\|}{P}}(OCH_3)_2$ | 70 |
| $C_3H_5IO_2$ | $ICH_2CH_2COOH$ | 112 |
| $C_3H_6BrCl$ | $ClCH_2CH_2CH_2Br$ | 20 |
| $C_3H_6O$ | $CH_3\overset{O}{\overset{\|}{C}}CH_3$ | 8 |
| $C_3H_6O_2S$ | $\begin{matrix}H & & H\\ & C{=}C & \\ H & & SO_2CH_3\end{matrix}$ | 100 |
| $C_3H_7BrO$ | $CH_3CHBrCH_2OH$<br>$CH_3CHOHCH_2Br$ | 50 |
| $C_3H_7NO$ | $H\overset{O}{\overset{\|}{C}}NHCH_2CH_3$ | 80, 82, 84 |
| $C_3H_8O$ | $\begin{matrix}CH_3\\ \|\\ H-COH\\ \|\\ CH_3\end{matrix}$ | 26, 28 |
| $C_4H_6BrN$ | $BrCH_2CH_2CH_2CN$ | 120 |
| $C_4H_6O_2$ | $CH_2{=}CHO\overset{O}{\overset{\|}{C}}CH_3$ | 100 |
| $C_4H_8BrCl$ | $ClCH_2\overset{CH_3}{\overset{\|}{C}}HCH_2Br$ | 30 |

| Molecular formula | Structure | Page number |
|---|---|---|
| $C_4H_8O$ | $CH_3C(=O)CH_2CH_3$ | 16 |
| $C_4H_8O_2$ | $O(CH_2-CH_2)_2O$ (ring) | 8 |
| $C_4H_{12}Si$ | $(CH_3)_4Si$ | 8 and most scans |
| $C_5H_3Cl_2N$ | 2,3-dichloropyridine (Cl, Cl, N ring) | 96 |
| | 2,6-dichloropyridine (Cl, N, Cl ring) | 98 |
| $C_5H_{11}NO$ | $HC(=O)N(CH_2CH_3)_2$ | 78 |
| $C_6H_8N_2S$ | $NCCH_2CH_2SCH_2CH_2CN$ | 112 |
| $C_6H_{11}N$ | $[H_2C{=}CH{-}CH_2]_2NH$ | 122 |
| $C_6H_{11}NOS$ | $CH_3C(=S)N(CH_2-CH_2)_2O$ (ring) | 144 |
| $C_6H_{12}$ | $CH_2-CH_2$ / $CH_2$ $CH_2$ / $CH_2-CH_2$ (ring) | 8 |

| Molecular formula | Structure | Page number |
|---|---|---|
| $C_8H_7BrO$ | $Br-C_6H_4-C(=O)-CH_3$ (as impurity) | 150 |
| $C_8H_8O$ | $C_6H_5-CH(O)CH_2$ (epoxide) | 100 |
| $C_8H_9NO$ | $C_6H_5-CH_2CONH_2$ | 86 |
| $C_8H_{10}N_2O$ | $(CH_3)_2N-C_6H_4-NO$ | 146 |
| $C_8H_{11}NO$ | $CH_3CH_2O-C_6H_4-NH_2$ | 112 |
| $C_8H_{18}O$ | $(CH_3)_2CH-CH_2-C(CH_3)_2-CH_2OH$ | 32 |
| $C_9H_8O$ | $C_6H_5-CH=CH-CHO$ | 108 |
| $C_9H_9NO_3$ | $CH_3CONH-C_6H_4-COOH$ | 136, 138 |

| Molecular formula | Structure | Page number |
|---|---|---|
| $C_9H_{10}O$ | H, $CH_2OH$, C=C, H | 124 |
| $C_9H_{11}N$ | $CH_2$, $CH_2$, NH, $CH_2$ | 130 |
| $C_9H_{12}O$ | $-CH_2OCH_2CH_3$ | 40 |
| $C_9H_{13}N$ | $NH_2$, $-CH_2CCH_3$, H | 156, 158 |
| $C_9H_{19}BrN_2$ | H, Br, C=C, H, $CH_2NHCH_2CH_2N$, $CH_2CH_3$, $CH_2CH_3$ | 38 |
| $C_{10}H_{10}F_4O_3S$ | F F O, $H-C-C-CH_2-O-S-$ … $-CH_3$, F F O | 126 |
| $C_{10}H_{11}BrO_2$ | H H, $H-C—C-H$, O O, Br— … $C-CH_3$ | 150 |

| Molecular formula | Structure | Page number |
| --- | --- | --- |
| $C_{10}H_{13}N$ | $CH_3$, NH | 132 |
| | NH, $CH_3$ | 128 |
| $C_{10}H_{13}NO$ | $CH_3O$, H, H, H, N, H | 100 |
| $C_{10}H_{16}O_2$ | $CH_3$, H, $O_2$, H, CH, $CH_3$ $CH_3$ | 92, 94 |
| $C_{11}H_{10}O$ | O, $CCH_2CH_2C{\equiv}CH$ | 134 |
| $C_{11}H_{14}O$ | $CH_2$, $CH_2$, $CH_2$, $CH_2$, $CH_3O$ | 54 |
| | $CH_2$, $CH_2$, CH, $CH_3O$, $CH_2$ | |

| Molecular formula | Structure | Page number |
| --- | --- | --- |
| $C_{12}H_{14}N_2O$ | O, N, N, $CH_2CH_3$, $CH_2CH_3$ | 22 |
| $C_{12}H_{20}O_2$ | H, $CH_3$, C=C, $CH_3$, $CH_3$, H, $CH_3$, H, $COOCH_2CH_3$ ; $CH_3$, H, $CH_3$, C=C, $CH_3$, H, H, $CH_3$, $COOCH_2CH_3$ | 62 |
| $C_{14}H_{12}O_2$ | H H, C–O, C, H OH | 152, 154 |
| $C_{14}H_{14}O$ | $-CH_2CHOH$ | 52 |
| $C_{14}H_{14}O_3$ | O H O, –C–C–COCH$_2$CH$_3$, $CH_2C{\equiv}CH$ | 196, 198 |
| $C_{14}H_{30}O$ | $CH_3(CH_2)_{12}CH_2OH$ | 10, 12 |

| Molecular formula | Structure | Page number |
|---|---|---|
| $C_{15}H_{13}Cl$ | 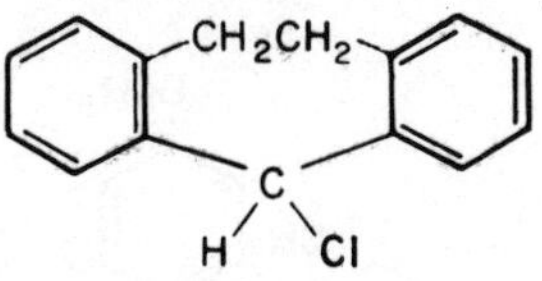 | 162 |
| $C_{16}H_{15}NO$ | 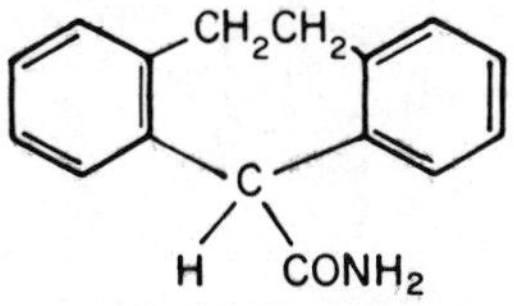 | 160 |
| $C_{16}H_{16}O_2S$ | 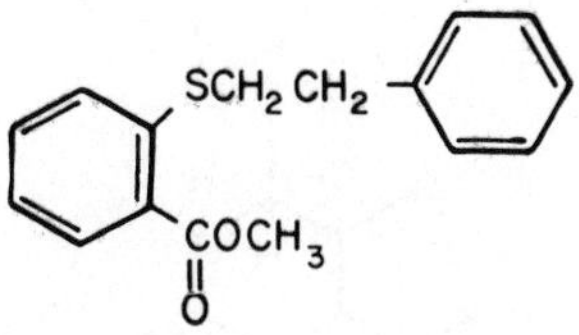 | 116 |
| $C_{16}H_{17}ClN_2O$ | 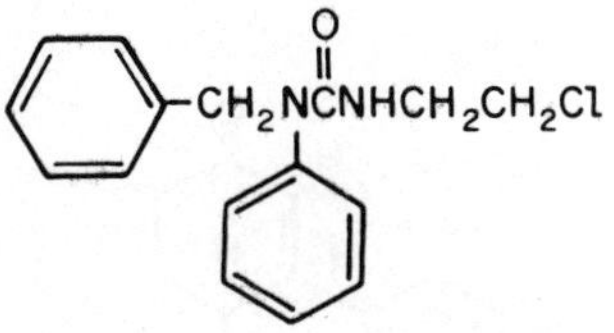 | 140 |
| $C_{17}H_{20}N_2O_2S$ | 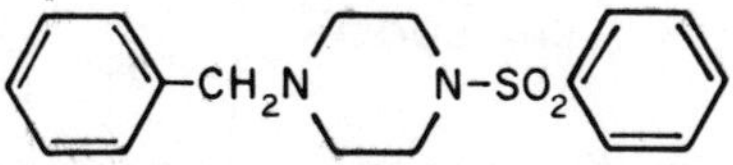 | 112 |
| $C_{17}H_{28}O$ | 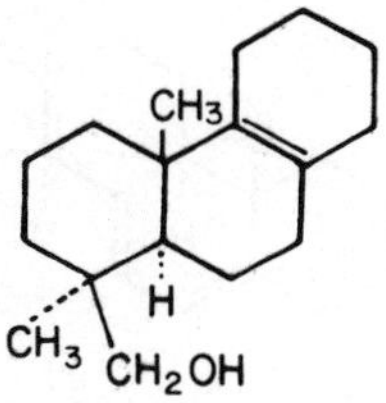 | 88 |

| Molecular formula | Structure | Page number |
|---|---|---|
| $C_{18}H_{20}O_4$ |  | 166 |
| $C_{18}H_{26}O_3$ | 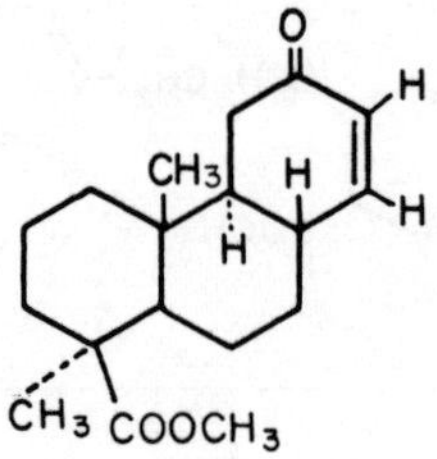 | 170,172 |
| | 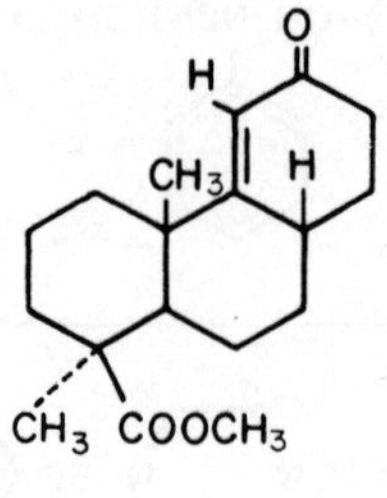 | 168 |
| $C_{19}H_{26}O_2$ | 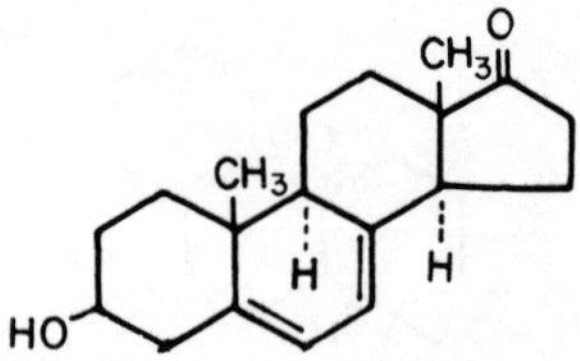 | 178 |

| Molecular formula | Structure | Page number |
| --- | --- | --- |
| $C_{20}H_{28}O_3$ | 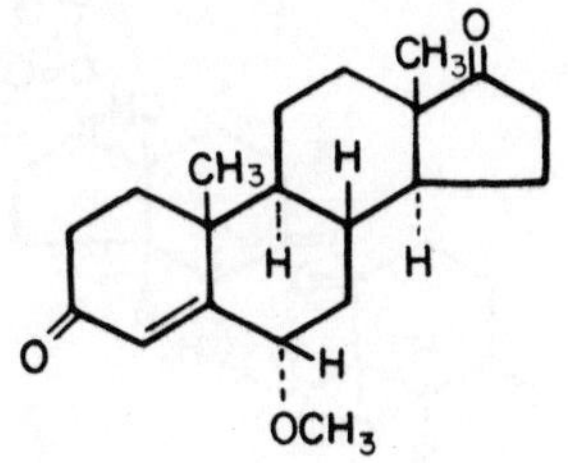 | 188, 192, 194 |
| | 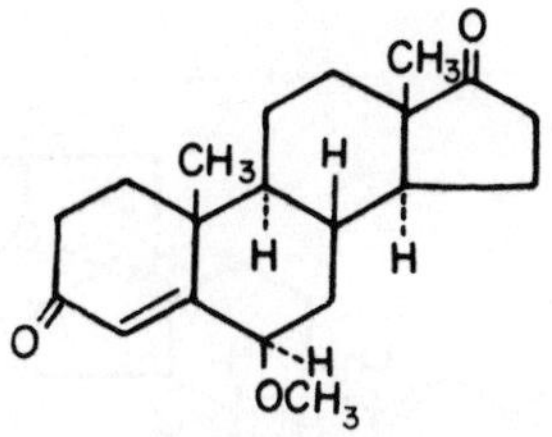 (in mixture) | 190 |
| $C_{21}H_{24}O_4$ | 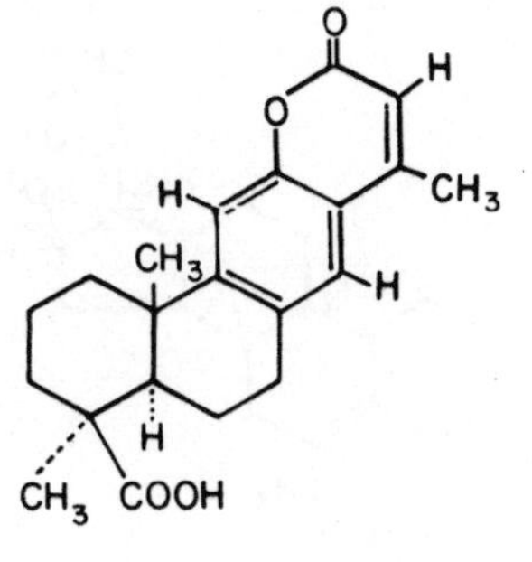 | 56, 58 |
| $C_{21}H_{27}BrO$ | 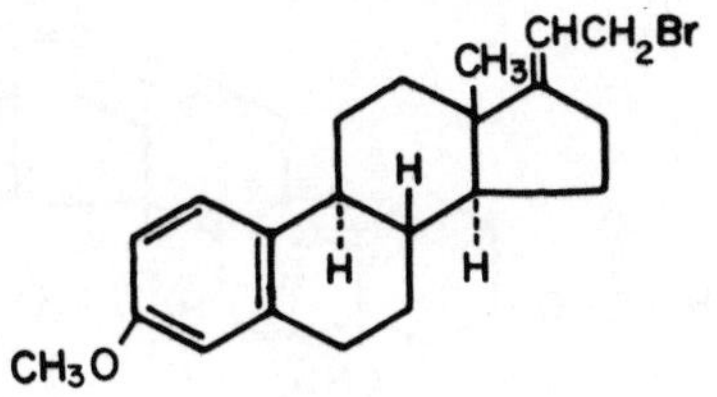 | 186 |

| Molecular formula | Structure | Page number |
|---|---|---|
| $C_{21}H_{29}ClO_4$ | 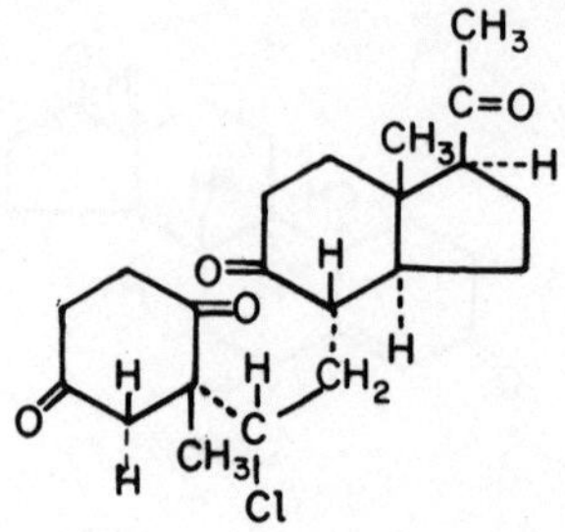 | 174 |
| $C_{22}H_{31}BrO_3$ | 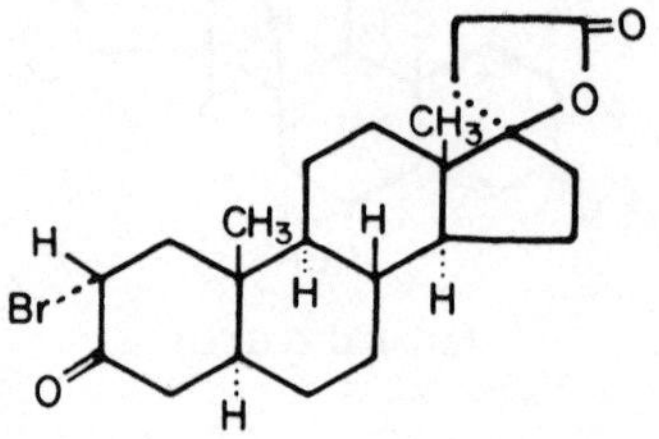 | 110 |
| $C_{22}H_{31}FO_3$ | 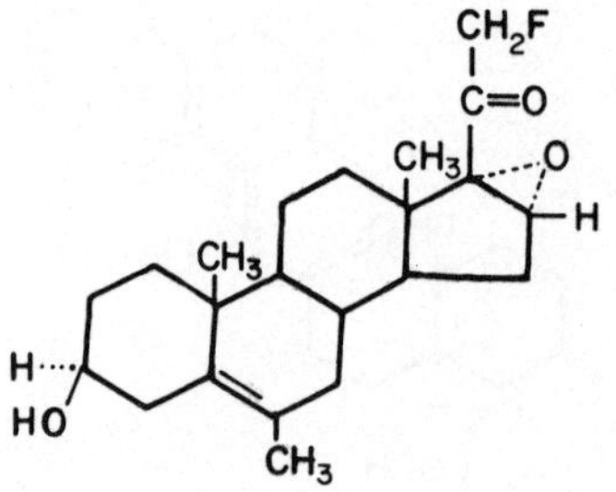 | 182, 184 |
| $C_{23}H_{33}ClO_5$ | 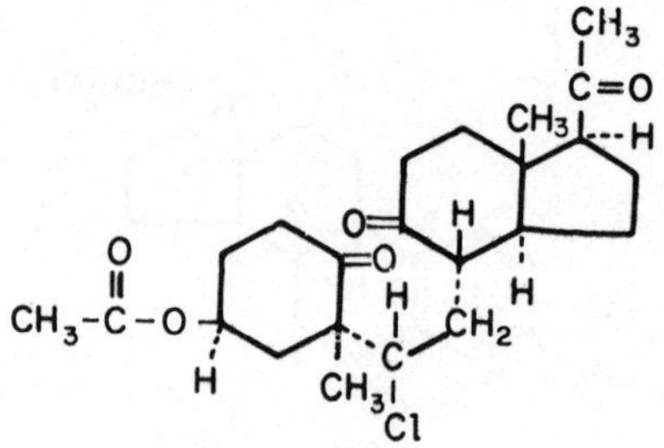 | 176 |

| Molecular formula | Structure | Page number |
|---|---|---|
| $C_{24}H_{30}O_4$ | 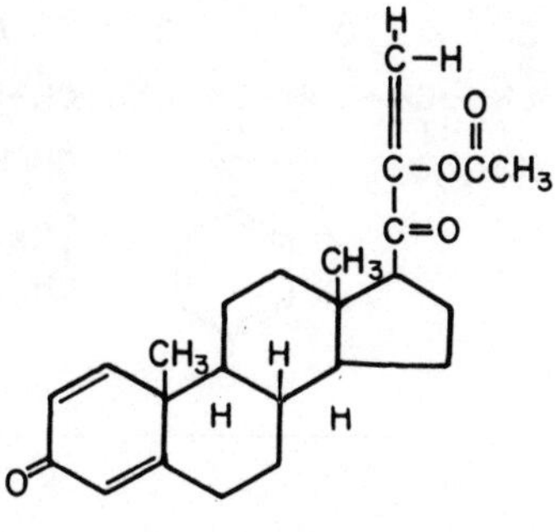 | 180 |
| $C_{24}H_{37}NO_3$ | 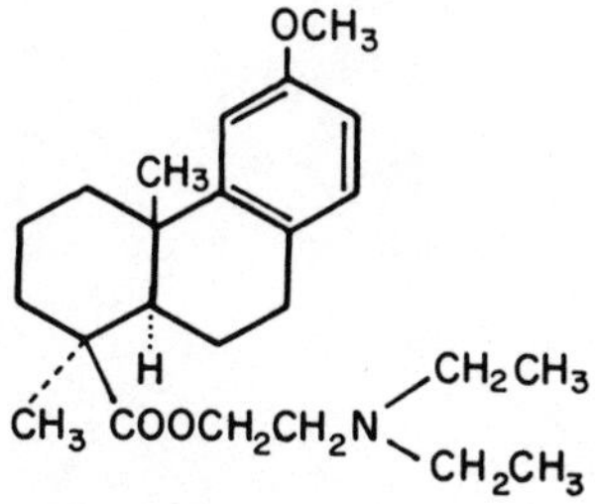 | 60 |
| $C_{27}H_{38}O_7$ | 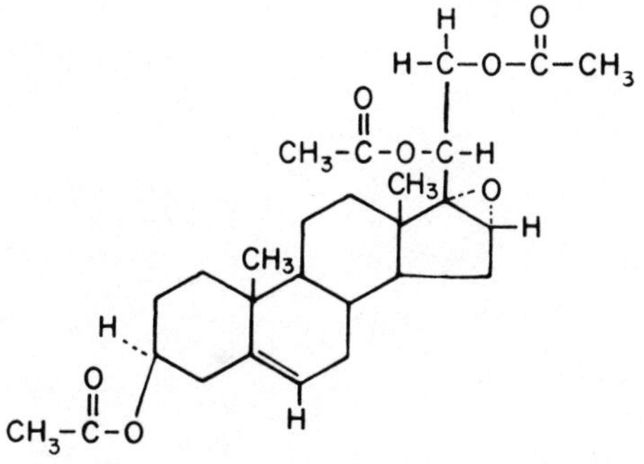 | 102, 104, 106 |

| Molecular formula | Structure | Page number |
|---|---|---|
| $C_{35}H_{46}N_6O_8$ | | 200 |

$$C_6H_5\text{-}CH_2O\text{-}\overset{O}{\overset{\|}{C}}NH\text{-}\underset{(CH_3)_2CH}{\underset{|}{CH}}\text{-}\overset{O}{\overset{\|}{C}}NH\text{-}\underset{CH_2C_6H_4OH}{\underset{|}{CH}}\text{-}\overset{O}{\overset{\|}{C}}NH\text{-}\underset{CHCH_3CH_2CH_3}{\underset{|}{CH}}\text{-}\overset{O}{\overset{\|}{C}}NH\text{-}\underset{CH_2\text{-imidazolyl}}{\underset{|}{CH}}\text{-}\overset{O}{\overset{\|}{C}}\text{-}OCH_3$$

# SUBJECT INDEX

**Page numbers cited refer to spectra which illustrate or introduce the topics listed.**

B



C

D



E

## M



## N



## O



## P



## Q



## R